# Writing on the Move

# Writing on the Move

## *Form, Practice and (Im)Mobility in Nineteenth to Twenty-First-Century Travel Writing*

Edited by

Samia Ounoughi and Tim Hannigan

berghahn

NEW YORK • OXFORD

www.berghahnbooks.com

First published in 2026 by
Berghahn Books
www.berghahnbooks.com



**Library of Congress Cataloging-in-Publication Data**

A C.I.P. cataloging record is available from the Library of Congress
Library of Congress Cataloging in Publication Control Number: 2025045261

**British Library Cataloguing in Publication Data**

A catalogue record for this book is available from the British Library

**EU GPSR Authorized Representative**

LOGOS EUROPE, 9 rue Nicolas Poussin, 17000, LA ROCHELLE, France
Email: Contact@logoseurope.eu

ISBN 978-1-83695-355-5 hardback
ISBN 978-1-83695-356-2 epub
ISBN 978-1-83695-357-9 web pdf

https://doi.org/10.3167/9781836953555

# Contents

# Illustrations

## Figures

## Table

# Acknowledgements

First and foremost, we wish to thank the contributors to this volume – both those who responded to our original call for papers, and those we convinced to come aboard once the project was underway. They have each drawn on their diverse specialisms to explore the concept of 'writing on the move' in all its nuances. We are particularly grateful for their reliably prompt and open-minded approach to the editorial process, which has made *Writing on the Move* into a properly collaborative project. We also wish to express sincere thanks to the anonymous peer reviewers, who went above and beyond the usual expectations in their detailed feedback on the manuscript and whose assistance has strengthened every section of this book. We are very grateful to Amanda Horn and Emma Wallace at Berghahn for their support and guidance throughout. We also wish to acknowledge La Société d'Etude de la Littérature de Voyage du Monde Anglophone (SELVA) for continuing to provide a collegial and supportive network for all those who study English-language travel writing.

# Introduction

## *Travel Writing and Movement*

Samia Ounoughi and Tim Hannigan

### Movement, Stasis and Travel Writing Studies on the Move

Travel writing might readily be described as a 'literature of movement'. Travel books are typically concerned with journeys, large, small, singular or multiple; and a travel writer's journey is, as Carl Thompson puts it, 'a movement through space' (2011: 9). Scholarship of travel writing has, naturally enough, often concerned itself with the mobility that the genre records, its movements both through space and between cultures. Pertinent to our focus here, Sidonie Smith (2001), for example, placed a particular emphasis on the way differing modes of travel – on foot, by bus, train or aeroplane, or mounted on a camel – shaped late-nineteenth- and twentieth-century travel writing by women, and how the differing modes of bodily movement affected the performance of gender within the resultant texts. Some more recent studies have focused specifically on the physicality latent in textual accounts of journeys. Charles Forsdick, for example, has considered the significance of corporeality and technology in this 'most physical of all literary genres' (2016: 68), while Tim Youngs has pointed out that both journeys and their narratives 'are shaped by the mode of transport adopted' (2015: 145). Yet within all this is concealed a tension, one perhaps more readily acknowledged by practitioners than by scholars: as the travel writer and journalist Samanth Subramanian points out, the process of composition 'happens at your desk, it doesn't happen outside ... the travel has happened, but the travel *writing* is happening at your desk' (Hannigan 2021: 225). If we bear this in mind, the designation 'travel writing' itself is revealed as a highly unstable and contradictory compound: the first word is a synonym for movement while the second is redolent of bodily immobility, suggesting a figure hunched at a desk, all physical motion restricted to the passage of the hand holding the pen or the fingers over the keyboard. Travel writing thus appears, at a glance, to be the most oxymoronic of genres: the practice of writing (like the practice of reading) typically requires stasis; and yet what that writing emerges from – and frequently seeks to represent – is movement.

However, this is not the whole story. Subramanian is correct to point out that the ultimate composition of a long-form travel text for publication typically takes place in conditions of desk-bound immobility; but many, if not most, professional travel writers will work from rough notes, daily journals and other mnemonic devices (including visual images) collected in the field – that is, during the movement that they seek to recount in their published texts. Indeed, when it comes to shorter and more ephemeral forms that may attract the attention of travel writing scholars – unpublished diaries, letters and postcards to friends and family and so on – the composition is likely to have occurred *during* travel (though the writer will, typically, still have required a relatively flat and stable surface on which to compose their text). In some cases, for example Dervla Murphy's first published book, *Full Tilt* (1965), letters never originally intended for publication form the basis of the final iteration presented to a public readership. Even when more thoroughgoing post-travel editorial processes are involved, the static object that is the journey's final iteration – the book – contains within it not simply a retrospective narrative of movement, but also traces and fragments, often deeply encoded, of that movement itself. Any binary distinction between movement and stasis itself similarly begins to break down the more closely a journey – or its textual record – is examined.

Every journey inevitably contains moments when bodily movement ceases: a traveller must rest, sleep, eat – and perhaps also pause to *write*.[1] And the mode of transport also impacts the possibilities for writing on the move. A writer travelling in a vessel – be it a boat, train, bus or perhaps even a desk in a virtual travel experience (an issue addressed in Chapter 9) – may be able to transcribe the experience *during* movement. Other travel modes, such as cycling, driving or even walking, may be too demanding for writing in the moment – though it is worth noting the intriguing example of Pierre Hyacinthe Azaïs, as cited by Philippe Lejeune, who, in the early 1800s, developed the device of a small board strapped to a walking stick, which made 'writing while walking . . . singularly easy and comfortable' (Lejeune 2009: 125). It remains unclear, however, whether Azaïs ever truly wrote *while* walking, or whether – perhaps more likely – he still tended to come to a halt when he had words to get down, for all that his invention allowed him to work easily in the field without needing to find a level surface and somewhere to sit. A traveller aboard a large ship, however – a ship bound for the Antarctic, perhaps, as discussed in Chapter 7 of this collection – may not only be able comfortably to write, but even to rest, eat and sleep while never ceasing to be in a state of movement. Digital technologies, meanwhile, have multiplied recording modes, extending the conventionally textual understanding of 'travel writing' to a wider sense of 'travel narrative'. Kylie Cardell and Kate Douglas, for example, have recently sought to frame the travel 'selfie' as a new mode of travel writing, typically posted to social media accompanied by 'strategically worded text' (2018: 105), but with the visual image very much the dominant narrative component. Such new technologies have also vastly expanded the possibilities for recording journeys in the moment. A mountain-climbing or kayaking writer equipped with a GoPro-type device may now be able to make a detailed and enduring record *during* movement in ways never previously possible.

How, then, do the conditions in which the first textual (or otherwise recorded) antecedents of a written travel narrative are recorded affect that finished narrative? And how do the processes of writing during movement – and indeed the writerly intention to record – affect the journey itself? These are the key questions that concern us here, and that have informed our discussions of 'writing on the move' in recent years. These discussions began in earnest with the panel on 'writing on the move' that we convened for the 2020 European Society for the Study of English (ESSE) conference – which itself took place in unusual conditions, the global COVID-19 pandemic having brought a halt to the bodily movement usually associated with international academic colloquia (the conference was delayed for a year, until 2021, and then took place virtually instead of in Lyon as originally planned). That initial conversation has now developed into this collection, which will, we hope, inform future considerations of writing and movement.

From the concurrent physical and emotional journeys of Ella Maillart and Annemarie Schwarzenbach in the late 1930s to the instantaneous digital discourses of contemporary urban travellers, and from the challenges of writing in the extreme conditions of the Antarctic to those of updating guidebooks via Google Street View, the contributions to this book examine the aforementioned tensions between movement and stasis. They also explore the convergences and divergences between actual journeys and the production of their resultant texts; for, as Christine Montalbetti (1997: 1–10) points out, every travelogue results from the challenge of representing a referential experience, with the meta-discourse on the travel experience and on the narrative inscribed in the text itself. Our contributors also ask questions about the meaning of 'movement' – and therefore what counts as 'travel writing' – in an age of virtual journeying and enforced immobility. This, in turn, speaks to the recurrent question of whether travel writing studies, or indeed the term 'travel writing' itself (see Borm 2004 for a useful and influential discussion of terminology), should include or exclude fictional travel accounts. We do allow discussion of fictional or semi-fictional texts in this collection, with chapters discussing Fitzgerald's *Tender is the Night* (1934) and Kerouac's *On the Road* (1957), for both novels are to a greater or lesser extent autobiographical, and both are undoubtedly born of their authors' actual travel experiences (and in the case of Fitzgerald, a composition process that itself was partially conducted on the move, as Chapter 3 explains). More significantly for contemporary debates, the inclusion here (in Chapter 9) of desk-based digital explorations helps to shift the scholarly focus to the dichotomy between physical and virtual travel, opening potential new tracks in the epistemology of travel writing studies.

As outlined above, travel writing studies faces – and occasionally overlooks – issues less apparent in the study of some other genres, namely the complexity and multilayered nature of the production of travel texts. Thus, in our considerations of 'writing on the move' we have attempted to emphasize the idea of travel writing as the product of active (in multiple senses) writerly practice rather than merely as a static corpus. This continues a wider recent shift towards a practice emphasis, and a critical-creative hybridity, in the scholarship of travel writing, for example in Gary Fisher and David Robinson's *Travel Writing in an Age of Global Quarantine* (2021) and Ben Stubbs and Lee Mylne's *Re-Thinking Travel Writing: The Journey of a Genre* (2024). Several of our contributors – Hannigan, Hall, and Chicot-Feindouno, Mansfield and Stothard – draw directly on their own experiences of travel and travel writing practice.

The chapters collected here are arranged to create a broad chronological and thematic flow – a continuous 'movement' across the book, apt to its subject matter. We begin with a chapter that travels across our entire historical time frame, from the early nineteenth century to the present, charting the shifting patterns of movement and stasis in travel writing practice – and the shifting nature of travel writing's 'truth claim' – produced by the illustrations sometimes included in travel texts. We then proceed to three explorations of movement and writing in the twentieth century – with gender a significant issue in each discussion – before two considerations of twenty-first-century writing on the move – in the ice-bound spaces of Antarctica, and in the rural British landscapes traversed on foot by the so-called 'new nature writers'. We then return to travel writing practice and the significance of images, with a self-reflexive study of contemporary guidebook-writing and the increasing likelihood that guidebook research will involve no conventional bodily travel at all. Finally, we turn directly to the practicalities of 'writing on the move', with a detailed exploration of modes of movement and stasis developed by Charlie Mansfield and his colleagues Clarisse Chicot-Feindouno and Mark Stothard for travel writing as a research practice. In ending with this chapter – a self-reflexive critical analysis, a subjectively observational account of a particular place *and* an instructional guide for future travelling writers – we seek to emphasize the sense of continuous movement back and forth between practice and theory that has informed all these discussions. The conclusion, meanwhile, draws out the key themes and resonances that have emerged from these explorations, and signals potential directions for future critical journeys among mobile writers and travelling texts.

## Mapping the Journey: Chapter Themes, Focuses and Summaries

The opening chapter of this volume – Tim Hannigan's 'Figures in a Landscape: Movement, Stasis and the Travel Writer as Image Collector' – stems from the very practical observation that travelling implies movement while writing, painting and photography all require stasis. Writing and image production thus appear incompatible with travel in its most literal sense, and yet, for centuries, many travel accounts have contained images. Through an analytical diachronic panorama of the travel text–image interplay from the Middle Ages to the turn of the twenty-first century, Hannigan's contribution explores the evolution of these relations through history, and resorts to an interdisciplinary method in which history intersects with image and travel writing analyses. This allows him to shed light on tensions in early travel texts produced by the presence of paratextual work by cabinet travel illustrators who were themselves merely recipients of travel accounts rather than travellers themselves, literally drawing on experience. While this is typically a phenomenon of the fifteenth to seventeenth centuries, Hannigan observes its continuation into the nineteenth century in the works on South East Asia of Thomas Stamford Raffles (1817) and John Crawfurd (1820), illustrated by William Daniell. Travel and stasis being hermetically distributed between writers and illustrators indicates a blend of referentiality and imagination in the edited versions of many illustrated exploration travelogues, even though they may have been intended to convey actual knowledge. Hannigan pursues his examination of graphic production through the case study of the British explorer George Hayward, whose archive from the Himalayas in the 1860s contains both written and visual materials, including maps and watercolours. Hannigan explores Hayward's poetics of truthfulness as a balance between the verbal expression of frantic motion and graphic production (of both watercolours and mapping data) demanding (and thus encoding a record of) stasis. Hannigan finally turns his attention to twentieth-century travel writing in which traditional iconographic productions began to vanish from self-consciously 'literary' anglophone travel books – a pattern that he connects to a shift from literality to literarity in the work of writers such as Colin Thubron, a changing emphasis from the accurate (re)presentation of a travelled place to the aestheticization of the traveller's experience.

As with Hannigan's case study of George Hayward, Chapter 2 also deploys a blended scholarly approach and a focus on nineteenth-century journeys. But here the setting is the watercourses of Europe, rather than the high mountains of Asia. And Kévin Cristin's 'Canoeing on the Waterways of Europe: Defining a Practice and a Genre through Robert Louis Stevenson's *An Inland Voyage* and Accounts of Other Travellers' also draws our attention to the significance of the specific means of transport – in this instance, single-seat canoes on the canals and rivers of Belgium and France – to the resultant texts. In his case study of Stevenson's *Inland Voyage* (1878), Cristin's method combines history, comparative literature and discourse analysis to shed light on the various contextual and textual factors that characterize Stevenson's text as innovative in the subgenre of river travel writing. Cristin compares *An Inland Voyage* with other canoeing travelogues of the same period, especially John MacGregor's *A Thousand Miles in the Rob Roy Canoe* (1866) and Philip Gilbert Hamerton's *The Unknown River* (1871). He demonstrates that Stevenson's *Inland Voyage* focuses more on the poetics of movement rather than on the depiction of landscapes. Cristin also examines Stevenson's correspondence and therein discovers evidence of his conviction that the essence of travelling and of travel writing is not the travelled place but motion itself. Cristin then examines the textual material through the lens of discourse analysis to highlight that canoeing shapes the whole structure of the travelogue as much as it shapes the poetics of movement. Syntactic and lexical analyses shed light on the multi-scalar movements of the author's body, his vessel and the river, all encapsuled in the arrhythmic flow of the narrative, oscillating between speed and strain on the one hand and quiet and contemplation on the other – a subtler variation of the abrupt binary of movement and stasis identified by Hannigan in George Hayward's archive.

Robert Louis Stevenson was, of course, a novelist as well as a travel writer; and the parameters of this collection do not restrict the concept of 'writing on the move' to non-fiction travel texts. Our third chapter, Elizabeth Bouzonviller's 'F. Scott and Zelda Fitzgerald's Literary and Visual European Journeys: A "Carnival by the Sea", and Many "Quick Odysseys"', takes twentieth-century authors best known for their fiction and sheds new light on the significance of their travelling lives to their work. Bouzonviller scrutinizes the works of Zelda and F. Scott Fitzgerald in light of the paratextual documents retracing their travels to Europe (and in particular France) during the 1920s. Using the tools of intertextuality, comparative literature and discourse and image analysis, Bouzonviller explores a multimodal corpus of writings and pictures, which include Zelda's scrapbooks, the couple's diaries, their correspondence, press articles and many more documents besides. She explores the impacts of travel on Scott Fitzgerald's novels *The Great Gatsby* (1925) and *Tender is the Night* (1934) and on Zelda's paintings and writings, including *Save Me the Waltz* (1932). Bouzonviller thus draws attention to the Fitzgeralds' travel experiences as a source of inspiration, highlighting the geographical links between the places the Fitzgeralds travelled and the settings of their fictions, and also the common traits between the people they met and the characters they created. The stylistic and narratological analyses demonstrate that travel experiences are deeply imprinted on the poetics of movement in the Fitzgeralds' fictions.

Bouzonviller extends her meta-discursive analysis by examining American newspaper articles to show how the Fitzgeralds used the press to promote their image as modern artists by cultivating the art of being remarkably absent from the USA. This chapter goes beyond the influence of travel on the Fitzgeralds' work to consider the wider significance of European travel among anglophone modernist writers and artists during the 'Roaring Twenties'.

Moving forward from the carefree hedonism of expat life in France in the 1920s to an attempted escape from darkening political and personal circumstances in Europe at the end of the following decade, in Chapter 4, 'Terror, Pity, Love: The Trials and Tribulations of Annemarie Schwarzenbach and Ella Maillart on the Road to Kabul', Julia Szołtysek sheds new light on a much-studied dual journey and its resultant paired texts. As a hypernym of *travel*, *motion* is a pivotal concept since its lexical field contains the psychological dimension of *motivation*, as well as the aesthetic notion of *motif*. Through the comparative analysis of Ella Maillart's *The Cruel Way* (1947) and the collection of essays and articles by Annemarie Schwarzenbach, *All the Roads Are Open* (2000 in German, 2011 in English), Szołtysek explores the travellers' motivations for travelling and for writing about travel. Queer studies and feminist studies intersect in this chapter as Szołtysek gathers two works composed by authors who travelled together from Switzerland to Afghanistan on the eve of the Second World War. These women stood out as avant-garde and nonconformist, resisting the gender conventions of the time, and – in the case of Schwarzenbach in particular – testing conventions of both sexuality and gender itself. Their quest for freedom and their common resistance to fascism is imprinted on their journeys and their books. Szołtysek thus analyses how these authors made use of long-distance motor travel, typically considered a male activity at the time. She also shows that by travelling in remote places and writing about them, these women were able to debunk scornful orientalist attitudes. By combining meta-discursive and psychological analyses of the recurrent textual and paratextual motifs, Szołtysek demonstrates that autobiography, which is often embedded within referential travel writing, is utterly multidimensional. She brings out the major editorial, temporal and thematic differences between the two authors' travel writing. Both decentre themselves in their work, but in very different ways. Schwarzenbach orientates her texts towards what she sees around her; Maillart orientates hers towards the physical and psychological condition of her travel companion. *The Cruel Way* is conceived as a monograph while *All the Roads Are Open* is a collection of press articles and essays edited as a book. *All the Roads Are Open* was written on the move and is a collection of depictions and reflections *in situ* while *The Cruel Way* is a retrospective (re)composition based on Maillart's notes and memories – providing a striking multivocal example of the movement–stasis dichotomy discussed in earlier chapters.

Maintaining an emphasis on motor transport, and on the movement–stasis interplay, as this collection moves forward through the twentieth century, Chapter 5 – Martin Wable's 'The Novel at the End of the Road: The Soul Circuit of Jack Kerouac as a Professional Writer' – brings a fresh perspective to an iconic text. Jack Kerouac's *On the Road* (1957), despite its notional presentation as a novel, is prominent among the mid-twentieth-century anglophone literature of travel, and because of its widely acknowledged autobiographical underpinning, comes close to matching influential scholarly definitions of travel writing such as Jan Borm's, mentioned above. Wable brings a valuable innovation to travel writing studies by making use of natural language processing (NLP) to analyse the alternance of movement and stasis in *On the Road*. Based on discourse corpus analysis, recurrent phrases featuring the construction 'end of' – such as 'end of the road' or 'end of the continent' – are identified and examined to bring forward the discursive rhythm of the narrative and, more broadly, the stakes of a running motif in a travel text. Wable shows the double semantic and narratological functions of the 'end of' motif. From a semantic point of view, the recurrence of 'end of' connotes the homodiegetic narrator's crumbled-down and scattered self as he is lost on the road. From a narratological perspective, the repetition of the 'end of' feature contributes to a narrative unity. Wable demonstrates that writing allows the main character to reconstruct a meaningful travel account out of a disorientated existence. The same method leads Wable to construe the 'end of' motif as a means to map the narrative structure and plot by delineating their spatial, emotional, aesthetic and social limits. Critical discourse analysis through NLP also fuels reflections on the various genres that coexist in *On The Road*. The text is an autofictional novel and a travel story that unevenly oscillates between referentiality and imagination, providing Kerouac much authorial freedom. In the diegesis, this freedom is expressed in the constant movement across the continent, back and forth, from east to west, yet entailing a constant evolution empowered by both travel and writing. Wable meanwhile observes that the travelogue also evolves, embracing features of various genres such as the comic strip, comedy, the picaresque novel and even epic. Just as the same journey is never the same experience for the traveller, the poetics that blend fiction and referentiality shape and reshape the travel text along the road, ever redefining the plastic contours of the travel writing genre.

Travelling does not automatically equate to freedom, and many works by female travel writers portray the challenges of travelling as a woman, particularly when doing so alone. These practical challenges may inform the creative poetics of the resultant travel accounts. In Chapter 6, '"A Strange Kind of Limbo": Navigating the Unknown in Contemporary Anglophone Female Travel Writing', Gemma Lake uses the concept of dynamic image related with affect to explore these specific traits in two referential travelogues about Africa displaying various travel modes and their stakes. *In Ethiopia with a Mule* (1968) by Dervla Murphy is an account of trekking through upland regions, and *Down the Nile: Alone in a Fisherman's Skiff* (2007) by Rosemary Mahoney is about a journey by boat. In order to show how the female travellers manage to navigate unguided, Lake uses the ethnological and anthropological concept of liminality (Van Gennep, Turner) as well as philosophy (Deleuze and Guattari). She thus brings forward the importance of corporeity and multisensorial experience that, in connection with affective perception, shape sensorial memories, the travelled environment (geographic and human) and the poetics of the travelogues. Lake analyses the expressions of effort and strain required in mountain trekking and rowing, manifestations of extreme travail that can lead to stasis. The analysis of corporeity also sheds light on gender issues like women using travel modes deemed specific to men (rowing in Mahoney's case) or unpracticable for either male or female foreign travellers (mountain travel on foot in Murphy's case). Furthermore, Lake observes that these travellers often faced clichéd perceptions of women's supposed insufficient physical strength or their inability to find their way without guidance. They challenged traditional conceptions of travelling by imposing their own modes of movement (Murphy) or concealing their gender (Mahoney). They consequently contributed to altering mentalities. Such strong female agency informs their poetics, which Lake analyses in the dynamic images that convey sensorial memories as well as individual and interpersonal affect.

Chapter 7 continues the emphasis on travel in extreme conditions. Kelly Hall's 'Writing from the Ice: An Examination of Contemporary Travel Writing in Antarctica' takes us to the end of the world, to 'the Ice' (Antarctica) where she personally worked at McMurdo Station during four austral summers in the early 2000s. As a remote and hostile environment with an extreme climate, the Ice is home to no one, meaning that it stands out as being exclusively inhabited by travellers, without permanently resident travellees. Yet, paradoxically, this chapter reveals that a visit to the Ice is more often typified by stasis than by movement. Hall is both a travel writing researcher and a traveller-writer on the move, and therefore she offers precious first-hand analyses of her experience. She also delves into texts written by her fellow travellers to debunk many myths about Antarctica as the epitome of solitude and danger, as it is often described in guidebooks and media representations. In her reflections on the stakes of travelling to the Ice, Hall draws a clear line between tourism and travel for work. Tourists seldom land on the continent and they are mainly catered for. This is different from people on a work mission who may live on the Ice for months in highly social stations where space is scarce. The Ice therefore profoundly reshapes its inhabitants because they are interdependent, and seldom get the chance to either explore the continent or simply be on their own. Hall's contribution also shows that the Ice experience transforms the traveller even before arrival. As a rare life experience, anyone wanting to go there will take any job, no matter their training or qualifications. Travelling to Antarctica thus implies being ready to be flexible and humble. It enhances the art of living together and the concept of belonging, hence the solid and long-term friendships the Ice has been the birthplace of. The traces of these precious bonds endure in what could be called Ice travel writing.

Chapter 8 returns to texts about less obviously hostile but more typically solitary travel experiences. Travel writing is about re-collecting memories; it is about 're/membering place(s)' (Delmas and Dodeman 2013). In 'Moving between Modes: Robert Macfarlane's and Kathleen Jamie's Journeys on Foot and in Time', Monika Kocot questions the genesis and conception of travelogues to find out how the travel writer's perception of the world and their evolution in life underpin the text's structure and poetics. She uses the geopoetic concept *Voyage-voyance* (travelling-seeing) in a double case study of Robert Macfarlane's *Underland* (2019) and Kathleen Jamie's *Surfacing* (2019). From landscape to wordscape, these two authors explore the (under)ground beneath their feet, each in contrasting yet complementary ways as both authors delve into their past travels to understand the current state of our planet. Kocot's analyses of the texts' structure, poetics and writing processes reveal that, as the semantics of their titles already suggests, *Underland* and *Surfacing* are reflections in an inverted mirror. Jamie's micro-observations show how small memories of past travels still physically resonate within her. This contrasts with Macfarlane's current and cataphoric observation of the underworld with a message of environmental emergency sent out to humans. Kocot observes that Macfarlane's journeys within the earth contrast with Jamie's focus on her intimate bodily experience of the world and her interactions with the other and the elsewhere. Kocot examines the pervasiveness of ellipses and silence in Jamie's text as opposed to the teeming intertextual references in Macfarlane's *Underworld*. The two writers' styles contrast in that Jamie shows while Macfarlane tells. Paratextual examination sheds light on the writing process and its consequences on the editing and poetics of the text. Jamie's rewriting based on notes taken decades ago translates into separate narratives, all characterized by the aesthetics of clustered momentum fragments. Macfarlane's ten-year work leads to a closely knit overall structure in which he accumulates experience and research, thus shaping the aesthetics of claustrophobia. That individual perception is the basic ingredient of travel writing seems obvious; Kocot's geopoetic analysis of *Surfacing* and *Underland* demonstrates that travel-related perception and memory may continue to evolve long after the travel experience – a form of psychological movement quite separate from the bodily progress of the physical journey.

The potential separation of travel writing from conventional bodily movement is further explored in Chapter 9, Tim Hannigan's '"Did They Even Go

There?": Virtual Experience and Projected Journeys in Contemporary Travel Guidebooks'. Within the very diverse genre of travel writing, guidebooks represent a major subgenre that must combine the practicality of a manual with aesthetic description of place. They are also, typically, assumed to have emerged from direct authorial experience of the place described. Yet the image of the author as a wildly mobile globetrotter – as suggested by some guidebook logos – is sometimes inaccurate. Guidebook composition does not always equate with writing on the move, as Hannigan reveals. As a travel writer, Hannigan has written guidebooks, and as an academic he specializes in travel writing analysis. Through this double lens, his contribution first offers a reflexive overview of guidebooks as a subgenre of travel writing. In this chapter, history and literary analysis intersect to put into perspective the widespread idea that guidebooks lack literarity and are utterly referential. Hannigan shows that guidebooks are not only informational; they must convey a sense of place, of genuine experience, a sense of atmosphere. As a result, literarity is identifiable within the guidebook (sub)genre. Hannigan also takes us behind the scenes of guidebook writing and revision throughout the last two decades. He highlights the impact of the COVID-19 pandemic and explains how the lockdown has increased the proportion of desk-based data collection over travelling in the field, especially for periodic updates of existing guidebooks. The author sits at their desk and watches videos, checks opening times, prices and so on, without necessarily ever visiting the destination in person. In observing this major evolution from within, Hannigan raises pivotal questions about motion and referentiality as key ingredients of travel writing. And yet, the semantics of bodily movement and the lexical field of travel still apply to research on the internet. Internauts use a *navigator*; they *visit* webpages as they *surf* the web. These displacements do not occur in a geophysical space, but in a virtual world. Does this invalidate the idea of travel? Is surfing the web not a mode of travel? Hannigan opens a new perspective on travel writing and referentiality, for although the guidebook author is at their desk, miles away from the place they are writing about, their experience is still real and the information still authentic.

Following on from the emphasis on travel writing practice in Chapter 9, the tenth and final chapter, 'Dialogue Journaling in Travel Writing Projects', is a self-reflexive analysis of travel writing in the making through the case study of a dialogue journaling experience in Plymouth (United Kingdom). Clarisse Chicot-Feindouno, Charles Mansfield and Mark Stothard present the creation and method of dialogue journaling as a team consisting of academics, writers and travel photographers. Dialogue journaling is a multimodal field practice inspired by sketchbooking, as well as an epistemological object for the three authors. Drawing conceptually on post-humanism, it is a rhizomatic writing process in which the authors immerse themselves in an urban space as part of its development as a potential tourist destination. The authors clearly classify dialogue journaling as a literary form of writing nourished by sound data collected on the ground. They carried out research, analysing a wide range of primary sources from historical local media to architectural archives. At the same time, dialogue journaling involves each team member's affect, creativity and imagination to shape, in their own style, both their journal and the future tourist spot. More than a travel writing subgenre, dialogue journaling also underpins a philosophical reflection on the mutual power of language and images on the one hand and space on the other. Through this project, the authors aim to identify the necessary ingredients and conditions that trigger what they call touremes, moments of interspatiality when the writer's immersion in a travelled place is the very setting of the text they are creating. Eventually the authors have regular meta-discursive exchanges on their work, which leads them to elaborate a discourse on the method for the use of future travel writers. Their reflection on intertextuality highlights the important role of fictions set in the emerging place they are touring and writing. They also tackle important practical choices to facilitate note-making while walking – bringing us back, once more, to the issue of an interplay of movement and stasis as a dominant aspect of travel writing practice – as well as the digital devices they used to adapt work on the ground and in their research laboratory. The three researchers show us that travel writing, which most often relies on memory, can also be cataphoric in mapping new places to be travelled and toured by future visitors.

**Samia Ounoughi** is a senior lecturer in English linguistics at Université Grenoble Alpes. She is a member of LIDILEM (Linguistique et Didactique des Langues Étrangères et Maternelles). She is also a member of LABEX ITTEM (Laboratoire d'Excellence Innovations et Transitions et Territoriales en Montagne) where she works with geographers, cartographers and historians. Her research deals with the relations between language and space, and she specializes in corpus discourse analysis of mountain travel writing. She co-edited 'Exceptions and Exceptionality in Travel Writing' (a special issue of *Studies in Travel Writing*, 2020) with Anne-Florence Quaireau. Her other publications tackle the process of naming features of mountain relief ('Referential Conventions as Compromise', in *Reference: From Conventions to Pragmatics*, John Benjamins, 2023). She is the co-editor of *Twenty-First-Century Perspectives in British Travel Writing: Decentring Epistemologies* (Routledge, 2025) with Emmanuelle Peraldo and Anne-Florence Quaireau.

**Tim Hannigan** teaches writing and literature at the Atlantic Technological University Sligo, Ireland. He received his doctorate – based on a critical-creative investigation of contemporary travel writing – from the University of Leicester. He is also an experienced professional travel writer, and the author of several narrative history books with travel themes. His most recent book is *The Granite Kingdom: A Cornish Journey* (Head of Zeus, 2023). His research has been published in the *Journal of Commonwealth Literature*, *Journeys*, *Terrae Incognitae* and *Studies in Travel Writing*, and in various edited collections.

## Note

1. Philippe Lejeune notes the existence in the early nineteenth century of small and highly portable desks containing all the essential writing equipment, intended for travellers; but he also observes that even with such equipment, 'it was still necessary to perch somewhere' and that 'Even when traveling, people wrote while seated on a chair in front of a table or something that served that purpose' (2009: 123) – even in the field, writing typically demanded stasis.

## References

Borm, Jan. 2004. 'Defining Travel: On the Travel Book, Travel Writing and Terminology', in Glenn Hooper and Tim Youngs (eds), *Perspectives on Travel Writing*. Aldershot: Ashgate, pp. 13–26.

Cardell, Kylie, and Kate Douglas. 2018. 'Visualising Lives: "The Selfie" as Travel Writing', *Studies in Travel Writing* 22(1): 104–17.

Deleuze, Gilles, and Félix Guattari. 2013 (1980). *A Thousand Plateaus: Capitalism and Schizophrenia*, trans. B. Massumi. London: Bloomsbury.

Delmas, Catherine, and André Dodeman (eds). 2013. *Re/Membering Place*. Lausanne: Peter Lang.

Fisher, Gary, and David Robinson. 2021. *Travel Writing in an Age of Global Quarantine*. London: Anthem Press.

Forsdick, Charles. 2016. 'Travel and the Body: Corporeality, Speed and Technology', in Carl Thompson (ed.), *The Routledge Companion to Travel Writing*. Abingdon: Routledge, pp. 68–77.

Gennep, Arnold van. 1910. *De quelques rites de passage en Savoie*. Paris: Leroux, 1910.

———. 1969 (1906). *Rites de Passage*. Paris: Mouton.

Hannigan, Tim. 2021. *The Travel Writing Tribe: Journeys in Search of a Genre*. London: Hurst.

Lejeune, Philippe. 2009. *On Diary*, ed. Jeremy D. Popkin and Julie Rak. Honolulu: University of Hawai'i Press.

Lyons, Agnieszka, and Samia Ounoughi. 2020. 'Towards a Transhistorical Approach to Analysing Discourse about and in Motion', *Topics in English Linguistics* 105: 89–111.

Montalbetti, Christine. 1997. *Le Voyage, le monde et la bibliothèque*. Paris: Presses Universitaires de France.

Murphy, Dervla. 1965. *Full Tilt: Ireland to India with a Bicycle*. London: John Murray.

Smith, Sidonie. 2001. *Moving Lives: Twentieth Century Women's Travel Writing*. Minneapolis: University of Minnesota Press.

Stubbs, Ben, and Lee Mylne. 2024. *Re-Thinking Travel Writing: The Journey of a Genre*. Basingstoke: Palgrave Macmillan.

Thompson, Carl. 2011. *Travel Writing*. Abingdon: Routledge.

Turner, Victor W. 1964. 'Betwixt and Between: The Liminal Period in Rites de Passage', *Proceedings of the American Ethnological Society: Symposium on New Approaches to the Study of Religion*, pp. 4–20.

Youngs, Tim. 2015. '"Take Out Your Machine": Narratives of Early Motorcycle Travel', in Julia Kuehn and Paul Smethurst (eds), *New Directions in Travel Writing Studies*. Basingstoke: Palgrave Macmillan, pp. 145–60.

CHAPTER 1

# Figures in a Landscape

## *Movement, Stasis and the Travel Writer as Image Collector*

Tim Hannigan

### Introduction

On 25 October 1868, the British explorer George Hayward spent the night outdoors, a short way east of the Chang La, a 5,391-metre pass in Ladakh, close to the Tibetan border. It was an intensely cold camping spot for a man travelling without a tent – and for the small group of local porters he had recruited to accompany him. At 7 a.m. the next morning – as part of his daily surveying routine – Hayward recorded a temperature of –11°C. In his subsequent report to his sponsors at the Royal Geographical Society (RGS) in London, he noted that it had been 'most difficult to keep anything liquid without being lost. Everything froze at once and burst the bottles.' And more significantly, 'Trying to paint in watercolours was out of the question. Water, brushes, and colours all froze together, and the enamel on the tin paint-boxes cracked from the intense cold' (Hayward 1871: 44).

Watercolour painting may strike us today as a faintly whimsical practice, a pastime for retirees in the English countryside rather than an activity for a hardy explorer in sub-zero temperatures thousands of metres above sea level. It may also appear strange that Hayward, so determined to travel light that he decided to forego even a tent for a late-autumn traverse of the world's highest mountain ranges, should find space in his baggage for a paintbox and sketch pad. But for Hayward painting was no mere hobby; it was a key component of what he plainly saw as the essential professionalism of his journey, a means of producing visual data to complement his descriptive prose narrative and the mass of bearings, altitudes and temperatures he recorded. Crucially, watercolour painting was also the most practical available means of image-making for an explorer in the field. By the end of the nineteenth century, it would have been largely superseded by photography; but in the 1860s that practice was still dominated by heavy, plate-based technology, impractical for a small, fast-moving expedition like Hayward's. He had already produced images of lakes, mountains and Buddhist monasteries during his journey north from India through Kashmir and Ladakh; and six weeks later, in similarly extreme temperatures, he would somehow manage to stop the brushes and colours from freezing to record a rough sketch of the source of the Yarkand River, which he was the first person accurately to map.

Visual images have been associated with European travel narratives since the late medieval period, and travel-related imagery has received considerable scholarly attention. The idea of images – both painted and photographic – as a manifestation and component of colonial discourses is well established. Indeed, nineteenth-century French orientalist paintings – such as Jean-Léon Gérôme's *The Snake Charmer*, which has appeared on the cover of several editions of Edward Said's *Orientalism* (1978) – have become a kind of visual shorthand for the scholarly discussion of such discourses. Photography, too, as a component of travel and of tourism discourse, has received considerable attention, notably by Peter Osborne (2001). And yet, when it comes to the scholarship specifically of travel writing (as opposed to travel and tourism more generally), relatively little attention has been paid to the presence of visual images within primarily written texts. Stacy Burton's claim that the scholarship specifically of travel *writing* 'scarcely mentions, much less analyzes, the presence of photographs and visual images' (2014: 47) may be something of an overstatement; but it does reasonably draw attention to a general deficit – particularly when it comes to the potential of images to function as an element *within* the poetics of travel writing, rather than as mere decorative paratexts. One notable exception to this is Margaret Topping's discussion of Nicolas Bouvier's travel photography and 'the shifting relationships of complementarity, tension, or disjuncture between image and text' within his work, characterized by 'slippage, subversion and paradox' (Topping 2009: 318). Elsewhere, Stacy Burton herself provides a useful initial discussion of photography and travel writing; and more recently, a special edition of the journal *Studies in Travel Writing* on 'Travel Writing and the Visual', edited by Giorgia Alù and Sarah Patricia Hill (2018), brought a renewed focus on the discussion of images within travel writing studies. Still more recently, Charles Forsdick (2022) has significantly advanced that discussion with a focus on the role of photography in the twin 1930s travel narratives of Ella Maillart and Peter Fleming. Significantly, Forsdick draws specific attention to photography as part of the 'poetics of travel'.

This chapter aims to further develop the discussion of visual imagery and travel writing, with a brief survey of the changes in both image-making practices by travelling writers and the deployment of the resultant sketches and photographs as part of the rhetoric and poetics of the travel text, between the nineteenth and twenty-first centuries. It also seeks – in the context of this collection's overarching themes – to draw attention to the connection between image-making and *movement*. Travel writing is, almost always, an account of bodily progress through a landscape; it is 'the most physical of all literary genres' (Forsdick 2016: 68). And yet image-making – be it an attempt to paint a watercolour or record mapping data in sub-zero temperatures, or the framing of a landscape through the viewfinder of a camera – has, for most of the past two centuries, demanded *stasis*, even if only momentary.

I begin with a very brief account of illustrated European travel texts prior to the nineteenth century. This survey is necessarily partial, and explicitly focused on anglophone texts, as it is intended to provide context for the subsequent examination of the interplay of movement and stasis generated by and latently recorded in the watercolours and maps produced in the late 1860s by George Hayward. The chapter then considers the significance of such images – and of later photographs – as part of travel writing's rhetorical 'truth claim' and, finally, makes some tentative observations on the changing patterns of illustration in late twentieth- and early twenty-first-century British travel writing.

## Entertainment to Empiricism: Images and Early Travel Writing

The images that sometimes accompanied medieval travelogues were often produced by professional artists who had *not* accompanied the writers on their journeys.[1] In the absence of direct experience (and resultant rough field sketches), we can reasonably assume that they based these images on the descriptions within the written text, on previous images purporting to represent similar scenes and locations – and on their own imaginations. The results may appear unintentionally comical today: the engravings representing South and South East Asia in early editions of the *Travels* of Marco Polo and John Mandeville usually feature obviously European costumes, landscapes and architecture. However, it is important to recognize the limited visual references possessed by these texts' original European audiences. A strictly accurate visual representation of India or Java might have been less readily receivable in terms of meaning for a fourteenth- or fifteenth-century readership than an image with familiar European elements. If asked to imagine a palace, a market or a king in a place called 'India', they would presumably have turned, like the books' illustrators, to their local – and *only* – readily available reference points for what such things looked like. Conversely though not contradictorily (for such images typically still featured European-style costumes and so on), Alù and Hill observe that fifteenth-century travel book illustrators may have 'enhanced or exaggerated the exotic appeal of the places in order to meet patrons' demands for the marvellous, and encourage ambivalent attitudes towards exotic areas' (2018: 3). And while it is always difficult to make judgements about the original reception of premodern texts, it seems reasonable to assume the

**Figure 1.1.** An illustration from an early fifteenth-century edition of *Mandeville's Travels*. The image depicts the Middle Eastern ports of Jaffa and Tyre, but the architecture and costumes are more reminiscent of Central Europe, where it was likely produced. Wikimedia Commons, public domain.

existence of a latent understanding among medieval readers that such images were *not* produced during the journey itself, and were thus not, directly, the result of movement. In the case of Anton Sorg, the influential German illustrator of a 1482 edition of Mandeville's book, whose images were recopied in other editions for centuries to come (Bennett 1953: 60), the obvious inauthenticity is hardly surprising: the author of the text itself was a pseudonymous hoaxer.

The phenomenon of travel text illustrations produced by artists who had not actually taken part in the journey and who had no direct experience of the places depicted endured for a surprisingly long time – even after the practice of professional artists accompanying major voyages and expeditions had become widespread in the eighteenth century. The images published in the major texts resulting from British colonial activities in South East Asia during the Napoleonic Wars – Thomas Stanford Raffles's *The History of Java* (1817) and John Crawfurd's *History of the Indian Archipelago* (1820) – were prepared by William Daniell, a professional artist who had never been to the places these books described. In some cases – notably the ethnographic portraits – Daniell was able to use as models actual people from South East Asia who had accompanied returning colonial administrators to Britain. At other times, he produced modified copies of images originally made by Dutch artists. For his scenic backgrounds, it is likely that he also drew on his own experiences in India and Sri Lanka – places several thousand miles north-west of those he was representing. Daniell's South East Asian images might, then, be seen to prefigure concepts usually associated with late twentieth-century postmodernism – copies of copies of an actuality, or 'third-order simulations' as Jean Baudrillard

**Figure 1.2.** An image of a ruined temple at Prambanan in Java, Indonesia, produced by William Daniell to illustrate T.S. Raffles's *The History of Java*. Daniell had never visited Java, and based this image on an earlier painting by the Dutch surveyor H.C. Cornelius. Public domain.

(1983) might describe them, in which any underpinning 'reality' is fragmented and in danger of being entirely lost.

Despite such examples, by the early nineteenth century, the idea that the illustrations found within anglophone travel texts might themselves be a product of actual travel was more firmly established. This can be attributed in part to the emphasis on sketching as an appropriate activity during aesthetically impelled domestic or intra-European travel. 'Sketching' had been a standard activity during the 'Grand Tours' of affluent Northern European youths since the seventeenth century; and in the eighteenth century William Gilpin, the British popularizer of the concept of 'the picturesque' – which he defined as 'a term expressive of that peculiar kind of beauty, which is agreeable in a picture' (1768: 2) – made extensive pedestrian journeys through Britain that he explicitly framed as 'sketching tours'. Watercolour painting or pencil sketching required only lightweight, portable equipment well suited to travellers moving at least partly on foot (as Gilpin tended to do), and by the turn of the nineteenth century 'making innumerable sketches' was a standard aspect of leisure itineraries in such destinations as the English Lake District (Buzard 2002: 46). William Daniell's own most substantial work was the eight-volume *A Voyage round Great Britain*, produced between 1814 and 1825 – an unusual piece of travel writing in which the written text is subordinate to the images – and in the same period, works of antiquarian travel on Greece and Italy sometimes featured illustrations produced by artists who had accompanied the authors on their journeys.

At the same time, the increasing professionalization of exploration – given a particular quasi-official imprimatur in Britain with the 1830 establishment of the RGS, which sponsored many prominent British explorers and provided the first public print outlet for accounts of their travels in its *Journal* and *Proceedings* – placed a heightened emphasis on formal data collection, including image-making. Larger nineteenth-century British expeditions often employed a dedicated professional image-maker – painters at the outset, photographers in later years.[2] In the case of a much smaller expedition, however, the principal explorer themselves might be expected to follow the example set by Gilpin, and to produce their own sketches at key points during their journey. This was the approach taken by George Hayward during his brief but remarkable career as an explorer.

## George Hayward: Movement and Stasis on the Road to Turkestan

George Hayward was born in Leeds, England, in 1839, the son of a land agent to the Earl of Cardigan. Following the early deaths of both his parents, he joined the British Army as an officer, aged 19, with a commission paid for by his father's former employer. He underwent initial training in Ireland, and was then posted to India; but after three years of apparently uneventful service, he resigned his commission in 1865. Details of his next three years are sketchy, but he seems to have remained in India and to have made a number of journeys in the mountains around Kashmir. Then, in early 1868, he travelled to London and presented himself to Henry Rawlinson, vice-president of the RGS, as 'desirous of active employment . . . on any exploratory expedition' (Keay 1996: 221), and was sponsored to return to Asia with the mission of exploring and mapping the Pamirs, the mountain range that dominates modern Tajikistan, and overlaps the current borders of Kyrgyzstan, Afghanistan, Pakistan and China. In the 1860s, this region was little known to Europeans, but it had considerable geopolitical significance, lying as it did in the narrowing space between the expanding colonial spheres of Britain and Russia.

Hayward initially attempted to reach the Pamirs via Ladakh and Eastern Turkestan (now the Chinese province of Xinjiang), mapping sections of the Karakoram and Kun Lun ranges along the way. However, the Eastern Turkestan authorities refused him permission to continue westward, and he was kept under house arrest in Yarkand and Kashgar for several months. On his eventual release, he returned briefly to British territory in India and made a fresh attempt – in the dead of winter – to reach the Pamirs via what was then known to Europeans as Dardistan and is now the Gilgit-Baltistan region of northern Pakistan. In the late 1860s, the army of the princely state of Kashmir (a signatory territory acknowledging British suzerainty, confected by the British and their ally Gulab Singh, the Raja of Jammu, in the 1840s) was engaged in a bout of aggressive territorial expansion in Dardistan, with the tacit approval of the British authorities. Having encountered evidence of atrocities committed by Kashmir troops in Dardistan, and having found the onward route to the Pamirs blocked by snow, Hayward returned once more to India and attempted – without success – to petition the colonial authorities on behalf of the people of Dardistan. He then made one final attempt to reach the Pamirs via the same Dardistan route, and was murdered in the Yasin Valley, west of Gilgit, on 18 July 1870 in circumstances that have never been satisfactorily explained.[3]

Because he died before completing his travels, Hayward's published outputs are limited: a series of redacted letters from the field and an abbreviated version of his formal report on the Eastern Turkestan journey variously published in the *Journal* (1871) and the *Proceedings* (1871) of the RGS. Much of the material produced during the later journeys in Dardistan is lost, and he never had the chance to write a book for a general audience. But the notes, journals, correspondence, surveying data and watercolours from the journey into Eastern Turkestan survive and are held in the RGS archives in London.

The overwhelming impression generated by an examination of Hayward's archive is of incessant *movement*. From his return to India in receipt of RGS sponsorship in the summer of 1868, he seems seldom to have spent a day longer than strictly necessary in one place, and his only periods of extended stasis were involuntary – when he was under house arrest in Eastern Turkestan, or when he was obliged to take time out from the road to compile reports and maps for his RGS sponsors. Departing on his initial journey from Murree in the Himalayan foothills on 26 August 1868, he made it to the Ladakhi capital, Leh, almost 700 kilometres to the north-east through mountainous terrain, in just twenty-seven days. In contrast, another British traveller, Robert Shaw, who coincidently headed into Eastern Turkestan concurrently with Hayward, took three months to reach Leh on a much shorter route via Kulu (Shaw 1871: 60, 71). Hayward's compulsion to move also manifests in individual moments. After an essential week-long stop in Leh to stock up on provisions, hire horses and recruit local porters (whose names, sadly though not untypically of the era, he never recorded), he was so eager to be underway that he departed at dusk rather than waste a single extra night in the town and covered the first 20 kilometres to Thiksey in darkness. At times, while deep in the mountains, he seems to have eschewed even sleep in his desire to maintain constant motion.

This apparent mania for forward motion is particularly evident in Hayward's account of a side-trip along the valley of the Yarkand River from the Eastern Turkestan frontier post at Shahidulla in the Kun Lun Mountains. Having been detained there by local officials, on 26 November 1868 Hayward abandoned most of his baggage and slipped away under cover of darkness accompanied by three Ladakhi porters and a yak, with provisions for just seven days (they later ate the yak). The following twenty days of extreme travel are recorded in detail in Hayward's official report to the RGS and an initial draft account, written up from rough notes in a small pocketbook while under house arrest several weeks later. Having crossed the 4,909-metre Kirgizjangal Pass from Shahidulla, Hayward travelled first downstream along the Yarkand River (which had never previously been surveyed by a European traveller, and which was marked inaccurately on existing British maps), and then upstream to its source. His movement-mania reached its apogee during a single thirty-six-hour period on 29 and 30 November, during which Hayward covered around fifty-five mountainous miles and went completely without sleep, at one point attempting to build a bridge across the river by moonlight:

> [A]s soon as a full moon rose above the mountains, and was shedding her silvery light far down the valley, we went on again down the left bank of the river for about nine miles, until stopped at a place where the stream runs deep and strong under a high bank to the left of the valley. *We wasted an hour trying to invent something on which to cross* [emphasis added]; but the long poles cut from the jungle close by, with

> which we endeavoured to form a temporary bridge, were washed away at once. Going back for a mile, we climbed the steep slope of the hill above the river, consisting of loose sand and shingle – the ground that gives way and lets one down about as fast as one progresses upwards. At length, descending again to the bank immediately above the river, we were arrested again a short distance further down, where a stream comes in from the south. This stream has carried into the main valley immense quantities of earth and debris, and now flows down out of sight between precipitous and over-topping banks, as if split by an earthquake. Steep heights enclose the valley on either hand, while above, to the north, rise the lofty snow-capped group of peaks in the Western Kuen Luen. *The valley, some 2,000 yards in width, was here found by observation of the temperature of boiling water to be 12,130 feet above sea-level* [emphasis added]. This was the furthest point down the Yarkand River which was reached. (1871a: 53–54)

This passage demonstrates Hayward's determined precision in the recording of detail. He gives a clear and precise account of his movements up and down the riverbank, and also includes sporadic flashes of aesthetically constructed prose ('silvery light'; 'lofty snow-capped group of peaks'), indicative of the way narrative travel writing and formal scientific reporting were not wholly differentiated in the mid-nineteenth century. The reference to the hour attempting to bridge the river as 'wasted' is symptomatic of the ferocious desire to maintain forward movement; but the second italicized line reveals a contradistinction. Hayward's bout of sustained motion along the Yarkand River was *not* without its pauses. Sleep, meal-breaks and bridge-building were a waste of time; but a halt to kindle a fire, boil water and record its temperature was worthwhile.[4] There were many such pauses throughout the journey.

Hayward recorded visual images in two ways: the aforementioned watercolours; and bearings, latitude readings and altitudes, initially taken down as numerical data and then later converted into a map. All of this required multiple pauses within the landscape, points where the physical journey ceased to progress. One such pause is directly recorded in the report of his explorations along the Yarkand River. Having reached the impasse on the downstream route described above, Hayward immediately turned back upstream, marching through the night. At dawn, he left his three Ladakhi companions and the footsore yak for a well-earned rest, and set out alone up the valley side:

> The steep slope of the mountain, covered with loose shingle and sand, was most unfavourable for climbing . . . After five hours' hard climbing, I reached the summit of the mountain in time to fix the latitude of the range by the sun's meridian altitude.

> The magnificent view which this station commanded was an ample reward for the toil of the ascent. Far away to the south and southwest stretched the high peaks and glaciers of the Karakoram and Muztagh Range, some of whose loftiest summits attain to the height of from 25,000 to 28,000 feet above the sea. One peak, situated to the east of the Muztagh Pass, reaches the stupendous elevation of 28,278 feet above sea-level, and is one of the highest mountains in the world. Beyond where the river sweeps out west the snowy peaks above the Kunjoot country were in sight towards Sarikol. East and west extended the whole chain of the Kuen Luen and the Kilian mountains, the last range to be crossed before the steppes and plains of Turkistan are reached, while immediately below lay the confined ravine up which the road ascends to the Yangi Pass, now full in sight beneath me. The extent of view of the main Karakoram or Muztagh chain comprised a length of 200 miles, stretching from near the Karakoram Pass to the head of the Tashkurgan territory north of Hunza and Nagar. (1871a: 54–55)

This moment is a striking example of what Mary Louise Pratt has termed the 'monarch-of-all-I-survey scene' in which Victorian travellers like Hayward developed 'a brand of verbal painting whose highest calling was to produce for the home audience the peak moments at which geographical "discoveries" were "won" for England' (2008: 198). Hayward was very much a colonial traveller, and his explorations have been framed as part of the 'Great Game' of geopolitical rivalry between Britain and Russia. He was explicit about the idea of 'winning' territory for the empire, at one point writing in a private letter to the RGS president of his hope that 'a Britisher may yet be the first to bag the [Pamirs]' (Hayward 1869a). However, it is the practice of his travel and image-making that is of particular interest here.

It is notable that Pratt refers to the monarch-of-all-I-survey scene as a form of verbal *painting*. But on that summit in the Kun Lun, Hayward was not only memorizing or noting down visual impressions for later textual or verbal recount; he may not have taken out his watercolours (the temperature in the shade was –15°C), but he boiled water once again to establish altitude and made multiple measurements with his sextant and artificial horizon, from which readings he later compiled the most detailed map of this section of the Greater Himalayan ranges then available. The summit he recorded at 28,278 feet (8,619 metres) turned out to be K2, surveyed from the north for the very first time, and his measurement – taken at a distance of more than 70 kilometres with portable equipment small enough to be carried in a leather belt – was inaccurate by just 27 feet (8 metres). He continued to take further measurements as he ascended the valley in the coming days. When he reached the source of the river on the afternoon of 8 December, he also managed to

**Figure 1.3.** George Hayward's watercolour of the source of the Yarkand River, painted in the field on 8 December 1868 in sub-zero temperatures. Courtesy of the Royal Geographical Society (with the Institute of British Geographers).

produce a rudimentary watercolour sketch, showing an icy plateau backed by glaciers and sharp peaks.

What George Hayward's example reveals is the intricate patterning of movement and stasis involved in any journey undertaken with a deliberate intention of recording (whether principally for a travel book, or for topographical or other scientific data). I do not here refer to the essential practical stops for rest and refreshment that are part of any extended journey, but instead to the moments of stasis explicitly prompted by the imperative to record. This includes the process of pausing merely to *look* at one's surroundings. For Pratt, the monarch-of-all-I survey scene features in imperial travel writing (and in contemporary travel writing with traces of latent coloniality) because 'the act of discovery itself... consisted of what in European culture counts as a purely passive experience – that of seeing' (2008: 198) and thus a dramatic climax needed to be retrospectively constructed in the resultant text. However, the process of seeing with the conscious intention of recording is *not* entirely passive: it involves actively committing one's impressions to memory or paper. And when actual image-making is involved – be it sketching, map-making or photography – the active nature of these moments of stasis-within-movement is heightened: measurements must be taken, temperatures recorded, photographic or sketching equipment set up. Indeed, it could reasonably be argued that such moments of image-making stasis are the most cognitively active parts of a journey. When the traveller is static, they may be in a state of more self-conscious intellectual engagement with their surroundings than when they are moving, particularly if the journey is undertaken on foot – where the focus during movement may more often be on one's own feet than on the landscape.

Discussing the travels of Maillart and Fleming in Eastern Turkestan seven decades after George Hayward, Charles Forsdick emphasizes the significance of photography to the 'poetics' of travel writing (2022: 98). But what we have here is the significance of image-making to the practice of travel itself, or to the poetics of travelling-to-write, perhaps. It is the impetus to record a visual image that determines the bodily movement – and periodic lack thereof – of the writer in the landscape.

## The Rhetorical Purposes of Image-Making

George Hayward's private account of the Eastern Turkestan journey recounts two alarming incidents in the uninhabited region north-east of Leh in November 1868: in the first he dropped his only sextant down a high mountainside; and in the second he discovered that he had mislaid the notebook in which he recorded his measurements. The first prompted 'utter despair for without a sextant how could I take latitudes?' and the second led to 'a nice to do, for if I did not find it my labours had been in vain' (Hayward 1869b). Happily, the sextant survived the fall, and the notebook was recovered; but his reactions suggest that Hayward felt that without the ability to survey, or with the loss of the data already collected, the entire journey would have been rendered worthless. Maps were clearly the paramount output of his explorations, and the production of new topographical knowledge for Britain his primary purpose. It is worth considering, then, why he bothered with his other image-making practice: the painting of watercolours. Hayward was, at best, a moderately proficient artist, and he certainly lacked William Gilpin's natural fluency with a brush (even making allowances for the generally more extreme environments that Hayward operated in).

The scientific value of his watercolours to the RGS was surely minimal when compared to his mapping data. Furthermore, not all of the images he painted during his travels could claim to represent 'new discoveries'. He recorded scenes in the Kashmir Valley, which was a well-established tourist destination by the 1860s, as well as sights along the road to Leh such as the Buddhist monastery at Lamayuru, which had frequently been recorded by previous European travellers. Given this, we might assume that for Hayward sketching was, after all, a mere whimsical pastime, something done for pleasure and quite out of keeping with the ascetic persona of a hardy professional explorer. However, I would argue that the watercolours served an important rhetorical purpose, underscored by the fact that he forwarded them to the RGS along with all his other materials: they were a key component of his 'truth claim' (Hulme and Youngs 2003: 3) as a traveller, for as Alù and Hill argue, 'visual images mark an experience as authentic; they confirm that objects, signs and elements . . . have been seen' (2018: 1).

As discussed above, the images found in medieval travelogues plainly did not represent the actuality of the journey described in the text, and we might reasonably conclude that their primary purpose was to *entertain* rather than to *inform*. Indeed, some scholars, such as Kirsten Sandrock, have argued that medieval travel writers themselves 'were still allowed and quite possibly expected to invent parts of their material' (2015: 189). But the post-'Enlightenment' emphasis on empiricism that governed Victorian travellers demanded rhetorical devices to reinforce the apparent veracity of their texts. Sara Mills identifies in travel writing from the late nineteenth and early twentieth centuries 'three main devices which are used to authenticate the text: foreground "factual" material; a strong narrator figure – both in terms of physical strength and knowledge – and photographic evidence' (1991: 136). As mentioned above, easily portable photographic equipment was not available to Hayward in the 1860s, but if we consider watercolours as a straightforward precursor to photography in this context (and it is worth reiterating that Hayward was not a notably skilled artist and would surely have gladly swapped paintbox for box camera had such a thing existed) then all three devices are clearly evident in the material he produced. He paused during his movement, long enough to record a reasonably accurate image of the landscape before him, to emphasize the fact that *he had been there*. The latent motivation for his images of the familiar scenes in Kashmir and along the Leh road now becomes apparent: they actually enhanced his truth claim, precisely *because* they could readily be verified by reference to the available images of the same places recorded by previous travellers. Thus, images of places that had *not* been previously recorded (and this includes both his maps and watercolours of places such as the source of the Yarkand) had their own individual truth claims reinforced.

Hayward also emphasized the direct experience-based principle in his correspondence. In the note to RGS president Roderick Murchison attached to his final map of Eastern Turkestan, he declared that 'No attempt, as you will observe, has been made to fill in the country situated at any distance from the line of the route which has not been personally explored and surveyed' (Hayward 1871b: 40). And in the unpublished first draft of his narrative of the journey, he made a number of scathing comments about a previous map-maker,

William Johnson, who had not been so assiduous and whose work Hayward found to be 'deplorably inaccurate' (Hayward 1869b).

The role of images as part of travel writing's rhetoric of truth-telling continued well beyond the nineteenth century – with photographs substantially replacing watercolours as the main mode of image-collection. Indeed, Mills (1991: 136) argues that the absence of photographs in Alexandra David-Néel's *Magic and Mystery in Tibet* – published in 1929 when photographic plates were increasingly ubiquitous in travel books – was one reason for the widespread scepticism with which the book was received (though it is probably fair to say that the notably fantastical elements of David-Néel's narrative were also significant here). Even in twenty-first-century travel texts, images retain a role in reinforcing travel writing's truth claim. Noting the inclusion of photographic plates in journalist Christina Lamb's *The Sowing Circles of Herat* (2002), Corinne Fowler observes that 'The result is significant: the travel book is presented as a text that carefully and accurately documents, rather than merely narrates, her experiences among Afghans. Overall, the effect is to raise the narrative's truth-status' (2007: 93). Despite this contemporary example, however, changing patterns can be identified in the use of photographic images in anglophone travel texts across the past century, with a possible shift away from an avowed adherence to 'truth-status' detectable in recent decades. The final section of this chapter will briefly discuss these shifts.

## Contemporary Travel Writing and Images: From Truth Claim to Aesthetic Status

By the turn of the twentieth century, photography had largely superseded the kind of watercolour sketching practised by George Hayward. Film and paper had replaced glass and metal plates in the 1880s, and the release of the mass-market Kodak Brownie camera in 1900 allowed for highly portable and simple photographic practice. This straightforward shift from one mode of image-making to another is nicely illustrated by the example of Isabella Bird. Her early publications typically featured engravings, presumably prepared from field sketches. Indeed, one of her 1860s publications was titled 'Pen and Pencil Sketches among the Outer Hebrides' (1866). But by the end of her career she had enthusiastically embraced photography. *The Yangtze Valley and Beyond* (1899) included 116 of her photographs – representing a multitude of moments of image-making stasis during her journey in China.

Photography became still more prominent in travel writing of the 1920s and 1930s, a fact 'exemplified by the growing visibility in travelogues of the period of affordable, portable photographic equipment' and 'epitomised by the popularity amongst travel writers of the Leica camera', according to Forsdick

THE AUTHOR'S BOAT.

**Figure 1.4.** One of the photographs, taken by the author, which illustrated Isabella Bird's *The Yangtze Valley and Beyond*. The boat is described in the text; the accompanying image appears to verify the narrative detail. Photograph by Isabella Bird, 1899, public domain.

(2022: 82). Forsdick also notes that photography could, by this stage, serve a practical mnemonic purpose for a travel writer, as well as forming part of the aesthetic of the published text. In the case of Ella Maillart, he writes, 'it is possible to discern an almost ekphrastic approach as *Forbidden Journey* draws on descriptions of the images she kept in front of her as she drafted her account' (ibid.: 98). We are here, arguably, entering Baudrillardian territory once more, for this would make Maillart's textual account of a scene in fact a description of a secondary image rather than of the original landscape – a copy of a copy. Forsdick also cites a line from Maillart's fellow traveller in Eastern Turkestan, Peter Fleming, particularly significant to the themes of this chapter: the portability and simplicity of their Leica cameras meant that 'a large proportion of the photographs we took were taken from the saddle; and it made a lot of difference being able to hold your horse in with one hand while you focussed the camera with the other' (cited in ibid.: 90). Fleming probably didn't mean that they took photos while actually riding along; rather that they were able to take a snapshot while briefly reining their horse to a halt. But we may here discern the first hint of an eventual erosion of the firm connection between image-making and stasis generated more recently by the ready availability of GoPro-type technology and the increasing contemporary ubiquity of 'point-of-view' footage of travel to be found on social media.

Photographic plates remained commonplace in travel books throughout the mid-twentieth century. These images generally remained subordinate to the text itself – a point that perhaps emphasizes the truth-signalling motivation for their inclusion. A grainy, unimaginatively framed black-and-white image of a building or landscape arguably added little to the aesthetic experience of the reader – and perhaps provided a less vivid evocation of the same place than the text itself; but it *did* confirm that the author had been where they said they'd been. Wilfred Thesiger is one of the few travel writers whose photographs are generally held in high regard as aesthetic objects in their own right. Discussing this, contemporary author and photographer Nick Danziger states that 'I cannot think of someone [other than Thesiger] who has produced both [text and images] to a [similar] level' and explains that he himself abandoned travel writing in the early 1990s to focus exclusively on photography because 'it was so intense that I couldn't actually do the writing and the photography at the same time' (Hannigan 2018: 222–23).

As photographic and printing technology advanced across the twentieth century, we might expect to find images becoming ever more ubiquitous in travel books. Surprisingly, this does not always appear to be the case – though Stacy Burton's claim of a noticeable mid-century shift in which 'photos all but disappear' from travel narratives (2014: 47) is perhaps an overstatement. They do, however, become noticeably less prominent in certain kinds of travel writing – in particular self-consciously 'literary' anglophone travel books. This shift is well illustrated by the five-decade travel writing career of Colin Thubron. His first travel books, *Mirror to Damascus* (1967) and *The Hills of Adonis* (1968), include conventional black-and-white photographic plates of the kind that had dominated travel books since the 1930s. His third book, *Jerusalem* (1969), was first published as a large-format hardback with lavish colour images (an atypical example in which the publisher *did* embrace technological advances), while his fourth book, *Journey into Cyprus* (1975), saw a return to an insert of black-and-white plates of places and people mentioned in the text. Thubron then took a hiatus from travel writing, and when he returned with *Among the Russians* in 1983, images were entirely absent, and they have made no return in any of his six subsequent travel books, up to *The Amur River* in 2021. Strikingly, the original images also vanish from more recent reissues of his first four books. A similar pattern can be partially identified in the oeuvres of some other travel writers with careers spanning the same period such as Geoffrey Moorehouse and Eric Newby, both of whose early books included photographs, but whose later works in the 1980s and 1990s generally didn't.

This shift can best be explained by the changes that occurred in the popularity and status of anglophone travel writing during Thubron's late-1970s hiatus. This period saw a remarkable flurry of high-profile publications, most significantly Paul Theroux's *The Great Railway Bazaar* (1975) and Bruce Chatwin's *In Patagonia* (1977), as well as Patrick Leigh Fermor's *A Time of Gifts* (1977) and Peter Matthiessen's *The Snow Leopard* (1978). These books were all markedly more self-conscious in their 'literariness' than the bulk of previous travel writing, and more concerned with the aesthetics of the text than with any strictly informational imperative. And few of them included photographs. During the 1980s, Colin Thubron was a key figure in an informal canonical cluster of anglophone travel writers, along with Chatwin, Theroux and Jonathan Raban, whose books were consistently imageless. This absence of photos most likely reflects an aspiration to heightened 'literary' status among the key practitioners of this period. Both Chatwin and Raban were, famously, uneasy with the designation 'travel writer', an anxiety likely related to that term's association with magazine journalism and guidebooks, and also to narrative travel writing's traditionally subordinate relationship to literary fiction. Eschewing photographic additions to the text could be seen as a disavowal of any lingering informational purpose: no one was likely to use *Among the Russians* as a practical guidebook. The removal of photographic plates also made travel books physically indistinguishable from novels – which very seldom include images.[5] But if photographs had long been a key prop of travel writing's truth claim, their abandonment suggests a subtle loosening, the pursuit of heightened literary status trumping the need to evidence veracity, and perhaps an erosion of the expectation – among practitioners if not their readers – that a travel book should be a strictly accurate recount of a journey.[6]

The pattern established in the late 1970s in which the most self-consciously 'literary' travel books did not include photographs continued through the 1990s and early 2000s, albeit with a number of exceptions. Books with an obvious journalistic emphasis, such as the aforementioned *The Sewing Circles of Herat*, were more likely to include photographic plates. Another atypical outlier is William Dalrymple. His first and third books (*In Xanadu*, 1989, and *From the Holy Mountain*, 1997) both included traditional black-and-white photographic plates. But Dalrymple, early in his career, was self-consciously writing in the tradition of the mid-twentieth century, with obvious tonal nods to the likes of Robert Byron, Peter Fleming and Eric Newby, and thus the inclusion of photo plates was in keeping with his own deliberately archaic posture. His more mature 2009 travel book, *Nine Lives*, does not include photographs.[7]

Perhaps unexpectedly, photographs do make a return to literary travel writing around the turn of the twenty-first century – but with a marked formal difference. Burton identifies this resurgence and notes that 'When visual images return to the genre at the turn of the millennium, they do so innovatively'

(2014: 48), citing the specific example of W.G. Sebald's *The Rings of Saturn* (1998). This is a book that, like the work of Bruce Chatwin before it, tests both the formal limits of travel writing and the conventional perimeter between fiction and non-fiction. As in Chatwin's *The Songlines* (1987), author, narrator and protagonist initially appear to be 'but one and the same' (Borm 2004: 17). But the absolute veracity of the individual elements of the narrative cannot be guaranteed. In keeping with the book's indeterminate status, the images are not held together in traditional glossy inserts, discrete from the body-text, and nor are they provided with captions; instead, they are scattered, without captions, throughout the text.

It may have been Sebald's example that established this particular formal mode of illustration, which has become increasingly common in more recent travel books. Similar inset images without captions feature in twenty-first-century works by, among many others, Kathleen Jamie, Philip Marsden and above all, Robert Macfarlane (see Chapter 8 of this collection). In Macfarlane's *The Wild Places* (2007), a single black-and-white photo appears at the start of each chapter. These images are all of landscape elements – treetops, icicles, a pathway, ripples in sand. But none is captioned, and with very few exceptions none of the locations would be identifiable, even by a person very familiar with the area in which the photo was taken. Each is plainly representative of the themes and landscape types explored in the chapter that follows – but the positioning means that this only becomes clear to the reader *after* they have initially viewed the image. And the lack of captions or direct reference points within the text means that any truth claim-reinforcing purpose is tenuous. Indeed, an examination of the book's publishing data page and acknowledgements section reveals that none of the photos were actually taken by Macfarlane himself: most were the work of John Beatty, with a few others taken by the author's parents (Macfarlane 2007: 330). In most cases, then, we can reasonably assume that these photographs have not emerged from the same journeys that the book describes, and may not even actually depict the same places. These, then, are photographs the sole purpose of which is aesthetic. The use of visual images to reinforce travel writing's truth claim has been abandoned here; and nor do the images any longer record and represent moments of authorial stasis amid the movement of the journey. In the literary travel writing of the twenty-first century we have in some ways returned to the mode of illustration to be found in the fantastical medieval *Travels* of John Mandeville.

**Tim Hannigan** teaches writing and literature at the Atlantic Technological University Sligo, Ireland. He received his doctorate – based on a critical-creative investigation of contemporary travel writing – from the University of Leicester. He is also an experienced professional travel writer, and the author of several narrative history books with travel themes. His most recent book is *The Granite Kingdom: A Cornish Journey* (Head of Zeus, 2023). His research has been published in the *Journal of Commonwealth Literature*, *Journeys*, *Terrae Incognitae* and *Studies in Travel Writing*, and in various edited collections.

## Notes

1. A noteworthy exception here is Erhard Reuwich, the fifteenth-century Dutch artist who illustrated Bernhard von Breydenbach's *Peregrinatio in Terram Sanctam* (1486), which 'is regarded as the first illustrated travel book' (Alù and Hill 2018: 3). Reuwich himself had made a pilgrimage to Jerusalem, an experience that informed his work for the book.
2. Ernest Shackleton employed both an official painter, George Marston, and an official photographer, Frank Hurley, during his ill-fated 1914 Imperial Trans-Antarctic Expedition. As well as still photographs, Hurley produced early moving images during his time in the Antarctic, making this a particularly well-documented expedition in terms of visual images.
3. For further detail on Hayward's travels and his murder, see Timmis (1998) and Hannigan (2011).
4. Measuring the temperature of boiling water was a relatively reliable method of ascertaining altitude for nineteenth-century explorers and surveyors: the temperature of boiling point falls as altitude increases.
5. Nineteenth-century novels did sometimes include imaginative illustrative engravings depicting scenes from the text. This convention lingered into the twentieth century, particularly in novels for children and young adults; but it seems now to have largely vanished.
6. The first edition of Chatwin's *In Patagonia* did include photos (though they are absent from some subsequent editions). Alù and Hill state that these serve to 'substantiate the narratives' truth claims by empowering them, through empirical validation' (2018: 11) – an interesting argument to make specifically about Chatwin, given his now well-established reputation for introducing a heavy element of fictionality into books likely to be received primarily as non-fiction travel texts.
7. Dalrymple's second book, *City of Djinns* (1993), is particularly unusual, in that it includes not photographs, but inset black-and-white watercolour images by his wife, the artist Olivia Fraser. A few of these watercolours represent specific people mentioned in the text, and are captioned; but most are not. This is thus a rare contemporary example of travel writing polyphony – or 'stereoscopic travel writing' (Forsdick 2009) – within a single book. It is not unprecedented historically, however. *Rambles Beyond Railways* (1851) recounts a journey undertaken by two people: Wilkie Collins and Henry Brandling. Collins wrote the text and Brandling painted the accompanying watercolours – a practice not entirely unusual in the nineteenth century.

## References

Alù, Giorgia, and Sarah Patricia Hill. 2018. 'The Travelling Eye: Reading the Visual in Travel Narratives', *Studies in Travel Writing* 22(1): 1–15.

Baudrillard, Jean. 1983. *Simulacra and Simulation*. Los Angeles: Semiotext(e).

Bennett, Josephine Waters. 1953. 'The Woodcut Illustrations in the English Editions of "Mandeville's Travels"', *Papers of the Bibliographical Society of America* 47(1): 59–69.
Bird, Isabella. 1866. 'Pen and Pencil Sketches among the Outer Hebrides', *The Leisure Hour: A Family Journal of Instruction and Recreation*, 646–50.
———. 1899. *The Yangtze Valley and Beyond*. London: John Murray.
Borm, Jan. 2004. 'Defining Travel: On the Travel Book, Travel Writing and Terminology', in Glen Hooper and Tim Youngs (eds), *Perspectives on Travel Writing*. Aldershot: Ashgate, pp. 13–26.
Burton, Stacy. 2014. *Travel Narrative and the Ends of Modernity*. Cambridge: Cambridge University Press.
Buzard, James. 2002. 'The Grand Tour and After (1660–1840)', in Peter Hulme and Tim Youngs (eds), *The Cambridge Companion to Travel Writing*. Cambridge: Cambridge University Press, pp. 37–52.
Chatwin, Bruce. 1977. *In Patagonia*. London: Jonathan Cape.
———. 1987. *The Songlines*. London: Jonathan Cape.
Collins, Wilkie. 1851. *Rambles Beyond Railways; Or, Notes in Cornwall taken A-Foot*. London: Richard Bentley.
Crawfurd, John. 1820. *History of the Indian Archipelago*. Edinburgh: Constable.
Dalrymple, William. 1989. *In Xanadu*. London: HarperCollins.
———. 1993. *City of Djinns*. London: HarperCollins.
———. 1997. *From the Holy Mountain*. London: HarperCollins.
———. 2009. *Nine Lives*. London: Bloomsbury.
Fleming, Peter. 1936. *News from Tartary*. London: Jonathan Cape.
Forsdick, Charles. 2009. 'Peter Fleming and Ella Maillart in China: Travel Writing as Stereoscopic and Polygraphic Form', *Studies in Travel Writing* 13(4): 293–303.
———. 2016. 'Travel and the Body: Corporeality, Speed and Technology', in Carl Thompson (ed.), *The Routledge Companion to Travel Writing*. Abingdon: Routledge, pp. 68–77.
———. 2022. 'Fleming, Maillart and Their Leicas: Photography and the Transformations of Interwar Travel Writing', in Barbara Korte and Anna Karina Sennefelder (eds), *Travel, Writing and the Media: Contemporary and Historical Perspectives*. Abingdon: Routledge, pp. 82–103.
Fowler, Corinne. 2007. *Chasing Tales: Travel Writing, Journalism and the History of British Ideas about Afghanistan*. Amsterdam: Rodopi.
Gilpin, William. 1768. *An Essay on Prints*. London: J. Robson.
Hannigan, Tim. 2011. *Murder in the Hindu Kush: George Hayward and the Great Game*. Stroud: History Press.
———. 2018. 'Interview with Nick Danziger', *Studies in Travel Writing* 22(2): 215–25.
Hayward, George. 1869a. Private Letter to Roderick Murchison. RGS archive item rgs812710.
———. 1869b. 'Journey from Leh to Shadula on the frontier of Eastern Turkestan with Exploration of the Source of the Karakash River – October and November 1868' (unpublished manuscript). RGS archive item rgs316845.
———. 1871a. 'Journey from Leh to Yarkand and Kashgar, and Exploration of the Sources of the Yarkand River', *Proceedings of the Royal Geographical Society of London* 14(1): 41–74.
———. 1871b. 'Letter from Mr. G. W. Hayward', *Proceedings of the Royal Geographical Society of London* 14(1): 40–41.

———. 1871c. 'Letters from Mr. G. W. Hayward on His Explorations in Gilgit and Yassin', *Journal of the Royal Geographical Society of London* 41: 1–46.
Hulme, Peter, and Tim Youngs. 2007. *English Association Issues in English 8: Talking about Travel Writing: A Conversation between Peter Hulme and Tim Youngs*. Leicester: University of Leicester.
Keay, John. 1996. *Explorers of the Western Himalayas, 1820–1895*. London: John Murray.
Lamb, Christina. 2002. *The Sewing Circles of Herat*. London: HarperCollins.
Leigh Fermor, Patrick. 1977. *A Time of Gifts*. London: John Murray.
Macfarlane, Robert. 2007. *The Wild Places*. London: Granta.
Maillart, Ella. 1937. *Forbidden Journey*. London: Heinemann.
Matthiessen, Peter. 1978. *The Snow Leopard*. New York: Viking.
Mills, Sara. 1991. *Discourses of Difference: An Analysis of Women's Travel Writing and Colonialism*. London: Routledge.
Osborne, Peter. 2001. *Travelling Light: Photography, Travel and Visual Culture*. Manchester: Manchester University Press.
Pratt, Marie Louise. 2008. *Imperial Eyes: Travel Writing and Transculturation*, 2nd edn. Abingdon: Routledge.
Raffles, Thomas Stamford. 1817. *The History of Java*. London: John Murray.
Said, Edward. 1978. *Orientalism*. New York: Pantheon.
Sandrock, Kirsten. 2015. 'Truth and Lying in Early Modern Travel Narratives: Coryat's Crudities, Lithgow's Totall Discourse and Generic Change', *European Journal of English Studies* 19(2): 189–203.
Sebald, W.G. 1998. *The Rings of Saturn*. London: New Directions.
Shaw, Robert. 1871. *Visits to High Tartary, Yarkand and Kashgar*. London: John Murray.
Theroux, Paul. 1975. *The Great Railway Bazaar*. London: Hamish Hamilton.
Thubron, Colin. 1967. *Mirror to Damascus*. London: Heinemann.
———. 1968. *The Hills of Adonis*. London: Heinemann.
———. 1969. *Jerusalem*. London: Heinemann.
———. 1975. *Journey into Cyprus*. London: Heinemann.
———. 1983. *Among the Russians*. London: Heinemann.
———. 2021. *The Amur River*. London: Chatto & Windus.
Timmis, Charles. 1998. 'George Hayward: His Central Asian Explorations, His Murder, His Legacy', *Asian Affairs* 29(2): 161–70.
Topping, Margaret. 2009. 'Phototextual Journeys: Nicolas Bouvier in Asia', *Studies in Travel Writing* 13(4): 317–34.

CHAPTER 2

# Canoeing on the Waterways of Europe

## *Defining a Practice and a Genre through Robert Louis Stevenson's* An Inland Voyage *and Accounts of Other Travellers*

Kévin Cristin

### Introduction

Prior to the expansion of the railways in the nineteenth century, European cities were already connected by a dense network of rivers and canals. In the early years of the Industrial Revolution, these routes had supported the increase of commercial activities in Britain and on the continent. A series of technological improvements progressively made locomotives and tracks more efficient, and more economically viable than barges. As a result, waterways were supplanted by the train system.[1] However, the decline of water transport created space for new uses of rivers and canals in the second half of the century. In Britain, boat races had already been organized on the Thames for more than fifty years. The 1830s had seen the development of rowing competitions at Oxford and Cambridge, and at Henley, with the creation of the Royal Regatta (Wenham 2020);[2] but the practice only concerned a relatively limited number of trained or amateur enthusiasts. In the context of the rise of leisure and tourism, waterways started being seen as a place for recreation. Fostered by the multiplication of rental companies, pleasure boating thus grew as an extremely popular activity (Wenham 2014).

The fashionable and easily accessible form of entertainment provided by these firms proved insufficient to those who were now looking for an original experience on the water. If popular pleasure boating only involved a limited amount of effort, travellers in search of a more challenging experience completed their journeys by rowing or paddling over long distances. In line with what James Buzard (1993) has labelled 'anti-tourism', many of these travellers also turned to distant, and what they perceived as marginal, locations. The waterways of Europe provided a fitting background for such expeditions.

In several instances, these personal adventures were turned into published accounts. The present chapter takes *An Inland Voyage* (1878), Robert Louis Stevenson's narrative of a canoe trip in Belgium and France in 1876, as a main point of entry into what represented a developing travel practice and literary genre during the period.[3] The primary focus on Stevenson will enable me to draw comparisons with the experience of earlier waterway travellers since the pioneers of the 1850s. Special attention will be dedicated to two other canoe narratives, respectively written by John MacGregor (1866) and Philip Gilbert Hamerton (1871).

In the second half of the nineteenth century, popular forms of tourism often defined travelling as going to a different *place* and enjoying *spending time* there. In 1873, due to his chronic lung condition, Stevenson was sent to southern France, which was then a common destination for medical tourism.[4] In Menton, when he was not confined to his sick room, he filled his days with walks in the neighbourhood and social occasions, attending fancy dinner parties with fellow residents. In contrast, with canoeing the value of the journey lay in the traveller's progression. In other words, movement – determined by the pilot and his embarkation as much as by the river and its flow – constituted an essential part of the trip. Thus, while this chapter traces the development of canoeing as a recreational activity, like other contributions to this volume, my examination will focus more specifically on the impact of mobility on narratives. This is where Stevenson's *An Inland Voyage* becomes particularly interesting. In some ways, the book follows already typical features of the travel genre; but the author also adopts a deeply reflexive attitude to his journey. More than an original or stimulating activity, canoeing appears as the fundamental prism of Stevenson's experience. As a result, the narrative often becomes *about* movement, with passages capturing the traveller's perception of his progression, and raising broader questions, especially about the pace of life in a changing world.

The present chapter contributes to the knowledge on the history and literature of travel, and to the scholarship on Stevenson. The culture that developed around the recreational uses of water in the nineteenth century has occupied an important place within histories of tourism as well as medicine. Previous studies have mainly focused on popular touristic practices in Britain and France through the development of institutions, like spas, and the emergence of popular sea locations, as places not only associated with pleasure but also with health (Anderson and Tabb 2002; Walton 1983, 2016). Barton and Brodie's anthology (2021) provides a vast array of primary sources on the practices that emerged in these locations. The coastal sea has also featured in histories of leisure, as the dedicated hotspot of yachting (Cusack 2000; Fisher 1997), while rivers have mostly been studied through the prism of sport practices like rowing (Wigglesworth 2013) or popular forms of pleasure boating (Wenham 2014, 2020).

Accounts of river navigation represent a relatively specific subgenre within the larger body of travel narratives of the period, and as such they have received little attention from scholars. The only existing works were written by a few enthusiasts. Paul A.L. Vine's account (1983) is an anthology of major narratives, with a substantial summary of each title and background information about the author and the journey. Although larger in scope, Mark Baldwin's two bibliographic surveys (Baldwin 1984; Baldwin and Burton 1984) also convey a particularly useful picture of the primary literature on canals, and Robert Burroughs (2019) has more recently provided a brief but valuable general survey of travel writing and rivers. As for Stevenson and his first travel narrative, despite its appearance in a monograph on the impact of travel on the author's works (Buckton 2007) and in a study on travel and time (Davies 2009), *An Inland Voyage* seems to suffer scholarly neglect due to its status as an early work by an aspiring writer.

I begin by tracing the development of canoeing in Europe as a recreational activity from which emerged a specific genre of travel writing. A brief overview of early narratives enables me to consider Stevenson's phenomenological picture of river navigation in *An Inland Voyage* as well as in his private letters from this period.[5] During his journey, the author did not only focus on his own perceptions; he also encountered other travellers, using different modes of transport, and reflected on their experience in comparison to his own. The second section explores how the practice of recreational rowing or paddling over long distances involved an attitude to travel that was fundamentally different from the one lying at the core of established forms of tourism. I argue that this new travel paradigm not only placed the focus on mobility itself but also on the active role of the body in this process. Associating recreation with adventure, this trend influenced the content and aesthetic of narratives. Finally, beyond moments of action, travelling on waterways also placed Stevenson in conditions that were favourable for contemplation. As his mind becomes attuned to the rhythm of the water, his reflections depart from his experience to touch on several aspects of life, on which he sometimes casts a particularly critical look. Taken together, these meditations come to define the spirit attached to the journey and constitute the author's personal contribution to the development of recreational outdoor adventures.

## Paddling Pioneers: MacGregor and Other Early Canoe-Travel Writers

Before Stevenson, relatively few individuals had completed a canoe journey in Europe and published an account of it. As one of the earliest means of transportation, canoes had long been used by Indigenous nations in several parts of the world. From the seventeenth to the late eighteenth century, they played a significant role in the colonial enterprise of the French and the British in North America (Erickson and Krotz 2021).[6] Later, in the first half of the nineteenth century, canoes are likely to have exalted the imagination of a broad audience in Britain, by featuring in the popular narratives of James Fenimore Cooper,[7] the American counterpart to Walter Scott.[8] They also appeared in the press and continued to do so throughout the century in the news reports on the British Empire. In these cases, the canoe remained represented as foreign and exotic, just like in the 1851 Great Exhibition in London, where models were displayed. However, ten years later, Salter Bros, one of the most prominent companies operating on the Thames, had fifteen canoes for rental (Vine 1983: xvi). The development of pleasure boating was thus instrumental in the adoption of the canoe as a popular recreational tool in Britain.

In 1865, John MacGregor was the first person known to have completed a long canoe journey for pleasure on the European continent. The following year, he published *A Thousand Miles in the Rob Roy Canoe* (1866) and founded the British Royal Canoe Club. He later took his canoe to Northern Europe (1867), then to the Eastern Mediterranean (1869), and produced extremely popular narratives.

MacGregor's first account follows the traveller's geographical trajectory from London to Germany, and along several European waterways, each chapter covering large portions of territory. As such, in the opening pages of the volume, the author feels compelled to define his enterprise by distinguishing it from more popular journeys and from the type of literature that flourished around them. Some areas of the region already attracted tourists, who would rely on steamers and railways during several-week-long tours. As a child, in 1863, Stevenson himself completed such a trip through European cities with his parents. To guide the crowds of British tourists, handbooks mapped the transportation network, suggested itineraries and provided a mine of practical information.[9] After quoting passages from the Baedeker and another similar volume, MacGregor acknowledges the value of such works, only to emphasize the originality of his own endeavour: 'This sort of guide-book is not to be ridiculed. It is useful for some travellers as a ruled copybook is of use to some writers' (1866: 7). Fashioning himself as both an experienced traveller and writer, MacGregor defines his journey as a singular adventure, located away from geographical and literary well-trodden paths.

An obvious difference between guidebooks and *A Thousand Miles in the Rob Roy Canoe* is that the latter depicts the journey in the form of a narrative. Although several sections were completed by carriage or railway – the canoe was purposely designed to fit into German wagons (MacGregor 1866: 8) – canoeing along waterways forms the unifying principle of the adventure.

Progress on land or water was punctuated by halts in villages, and the result is a collection of observations made on the way and captured on paper during these intervals. As a whole, the book records remarkable events and contains comments on navigation, landscape descriptions and reflections on the regions visited and their inhabitants. Interestingly, building his narrative from these various components, MacGregor often fails to convey a sense of the continuity of the voyage. As an early sentence suggests, his account rather turns into an accumulation of separate pieces: 'The voyage was chequered by innumerable little events' (ibid.: 22).

Only some passages tend to break this impression: 'These reflections are by the way; and we have been winding the while through a rich grassy plain till a bridge over the river made it seem quite a civilized spot, and, just as I passed under, there drove along one of the green-boughed waggons of jovial singers returning from Donaueschingen' (1866: 58). After dedicating several paragraphs to the specific pleasure associated with canoeing, the author resumes the narrative in a way that reproduces the experience of the moving traveller.

As we will see, MacGregor's journey and account drew inspiration from the model established by an earlier generation of travellers, who used a different type of watercraft. However, the Scotsman opened the way to followers and coined specific aspects of canoe narratives. Among the figures who preceded Stevenson was Warington Baden-Powell (1871). Unlike MacGregor, Baden-Powell took his canoe to open sea and made ample use of the front sail with which it was equipped. In his work, Stevenson only briefly records using the sail of his canoe early on in his journey, after which it became unnecessary. Finally, Philip Gilbert Hamerton (1871) completed a journey in France on a paper canoe of his own design.[10]

While the narratives of these travellers had driven the interest of readers, they left sufficient space in 1876 in the market of travel literature for Stevenson to imagine making an impression with his own adventure. When he started planning his journey, Stevenson was only at an early stage of his literary career and was in search of original material. Although he had already published short pieces in various magazines, including on travel, he still had not produced any full-length account, which was a necessary condition to gain recognition as a writer.[11] His correspondence a couple of weeks before departure shows how central this concern was:

> And I am glad I am off on my holidays. To the Highlands, first, to the Jenkins, then to Antwerp; thence, by canoe with Simpson, to Paris and Grez . . . to complete our cruise next spring (if we're all alive and jolly) by Loing and Loire, Saone and Rhone, to the Mediterranean. It should make a jolly book of gossip, I imagine. (Booth and Mehew 1994: 178)

Stevenson knew Walter Simpson from his student years at Edinburgh University. The two had already travelled several times together in France and Germany. In August 1874, they had spent a month on a yachting cruise in the Hebrides, in the west of Scotland, on board the *Heron*. Simpson and his brothers had themselves been keeping canoes for years in Granton and would often paddle in the Firth of Forth, as his sister later recalled. Sometimes, Stevenson would join his friend on the river on these occasions (Simpson 1912: 82–83).

The two started their journey in Antwerp and went towards Brussels, where they decided to take a train to avoid the long succession of locks lying on the way to Charleroi. They resumed their canoe trip at Maubeuge on the Sambre, and continued on the Oise until they reached Pontoise, their final destination. The second part of the cruise, to which Stevenson refers in his letter, was never carried out. Instead, he stayed for some weeks in an artists' colony in Grez, where he met his future wife, Fanny Van de Grift Osbourne, an American who had come to Europe with her children to study art. After his journey, the author did not go back to paddling until the next summer, when he returned to Grez. There, he only took his canoe out in the morning for a little bit of exercise on the Loing. After that, canoeing is mentioned again in 1879, in a letter where Stevenson refers to another cancelled journey, on the Forth this time, because of the weather (Booth and Mehew 1994: 318).

As mentioned in the introduction, Stevenson's account, *An Inland Voyage*, contains the same major ingredients as the works of his predecessors, but it proves original in the way movement becomes an object of consideration. To the word 'gossip', used in the letter quoted above, the author attributed the meaning of reflection on a particular topic.[12] In the case of his canoe journey, an important part of his thoughts concerned the prism introduced by the mode of transport on the experience of the traveller and its immediate impact on the narrative.

A few months before the publication of *An Inland Voyage*, Sidney Colvin, Stevenson's close friend and advisor, expressed concerns about the possible proximity between the narrative and James L. Molloy's *Our Autumn Holiday on French Rivers*, published four years earlier, in 1874. To Colvin, these two accounts of river navigation in the north of France risked covering the same geographical and literary territory, at the expense of Stevenson's volume . However, to Stevenson the boat made all the difference. In his reply to Colvin, the canoe stands at the basis of the aesthetic of his work: 'Molloy was not a canoeist; he went in a four oar gig. Moreover his book was not a book; it was only illustrated. Now my work is a true *Reisebild*; good or bad, well done or ill, it is the history of a man's life during some weeks of travel' (Booth and Mehew 1994: 231–32).

Opposing the sole figure of the canoeist to the boat used by Molloy and his companions, Stevenson points out that his single-seat canoe produced material conditions that were prone to provide the traveller with an individual experience of the journey, with direct influence on the content of the narrative. The kind of sociability emerging from a boat allowing space for three to five passengers indeed made travelling a collective adventure. Molloy's account opens on the conversation of a band of enthusiastic boys planning their next vacation: 'It was near the end of the season, and a few men just up from circuit had dropped in, and were discussing projects for the Long Vacation. Some were for the moors – others the Mediterranean. Ashantee was talked of – in fact every possible country and way of doing it' (Molloy 1874: 1). The rest of the narrative continues in the same joyful spirit and focuses mainly on the group's life as they move from place to place. On the contrary, like in other accounts of canoe trips, in *An Inland Voyage* Stevenson's companion often disappears as the narrative becomes the 'history of a man's life' on the water.

Molloy's adventure followed the earlier tradition of journeys from which MacGregor had himself drawn inspiration. In 1851, Robert Mansfield had completed his cruise in Germany in a four-oared gig and had published his account of it the following year (Mansfield 1852). After that, he had chosen a pair-oar for his new adventure on the Danube (Mansfield 1853). Following his example, Edmund Harvey (1854) had travelled through France, Belgium and Germany on a boat of similar design.[13] The two gentlemen had respectively been educated at Oxford and Cambridge, where they were exposed to the sporting culture that inspired their journey. At University College, Mansfield had practised rowing and had even competed against Cambridge in 1843 at the Grand Challenge Cup at Henley.

At the beginning of his own narrative, John MacGregor acknowledges some of these precedents, but quickly comes to underline other practical advantages of the canoe in his opinion: 'these very things, which are obstacles or dangers to the "pair oar," [*sic*] become interesting features to the voyager in a covered canoe.' MacGregor's canoe provided access to narrow or shallow parts of the river, allowing him to cover more ground. The traveller was also in a better position to appreciate his environment: 'For now, as he sits in his little bark, he looks forward, and not backward' (1866: 2). While MacGregor's main concern remained his experience of the journey, Stevenson was even more sensitive to the effect of his mode of transport on his literary material.

In one of the conversations transposed in his narrative, the author refers to another of his predecessors on the rivers. A few months before him, William Moens (1876) had followed a rather similar itinerary between Belgium and France, yet he was travelling on a steamer. During his stay at Grez in 1877, Stevenson and his friends started planning for a cruise on the canals of Europe on a converted barge, in the fashion of conventional pleasure journeys; but the whole scheme was eventually abandoned, and the barge was sold (Booth and Mehew 1994: 219 n. 2). If the author does not provide any further comment on Moens's journey in *An Inland Voyage*, other passages in his account show how he perceived the influence of this kind of bigger boat compared to the experience provided by his canoe.

Coming across other boats was part of the entertainment that Stevenson enjoyed on the river. Very early on in the narrative, he proves particularly interested in the barges that overtake his small canoe. The repeated sighting first leads him to contemplate the aesthetic beauty of these boats before considering the type of travel experience that they provide. In such passages, the author offers an acute phenomenology of another mode of transport. As he reflects upon the comfortable abode provided by the barge and the daily habits that emerge from life on board, he remarks that such boats allow one to combine the experience of travel with that of being at home:

> The chimney smokes for dinner as you go along; the banks of the canal slowly unroll their scenery to contemplative eyes; the barge floats by great forests and through great cities with their public buildings and their lamps at night; and for the bargee, in his floating home, 'travelling abed' it is merely as if he were listening to another man's story or turning the leaves of a picture book in which he had no concern. He may take his afternoon walk in some foreign country on the banks of the canal, and then come home to dinner at his own fire-side. (Stevenson 1878: 13)

As Stevenson comes to consider the specific visual perception of the landscape that surrounds the traveller on a barge, he immediately proposes to interpret it in narrative terms. At first, the linear motion of the landscape is designated as unrolling scenery, with the metaphor conveying the image of an unfolding parchment. The gesture associated with reading is then illustrated through another medium, that of the picture book. Alongside these visual tools, the formation of the images in the individual's mind is also considered through the act of listening to a tale. Despite these various sensory stimulations, the general impression conveyed by Stevenson through these images is that of a separation between the traveller and the world around him. Here, the author is thinking from the perspective of the experience provided by his own mode of transport, through which he was on the contrary looking for immediate sensations.

## Motion, the Body and Masculinity

Stevenson provides a sense of the connection he formed with his environment through the multiple descriptions of the changing landscape and weather during the entire journey; but this contact is far from always being harmonious.

If travelling on a pleasure boat could represent a relatively comfortable and enjoyable activity, Stevenson's enterprise exposed him to harsh conditions. The rain comes into the landscape of the narrative on multiple occasions, and the author is sometimes forced to confess its impact on his mood: 'And the rest of the time the rain kept coming in squirts and the wind in squalls, until one's heart grew weary of such fitful, scolding weather' (1878: 78). Because of their intensity and their negative effect, such conditions are depicted as inconvenient, but most of the time the rain is treated as an element that is part of the kind of experience pursued by Stevenson. The weather represents an example of the broader place occupied by accidents and challenges in the journey. Far from being avoided, occasionally difficult circumstances constitute one of the elements that contribute to making it exciting. Canoeing was not the only activity in which travellers were actively looking for such stimulation. It joined other growing practices in the second half of the nineteenth century, like mountaineering, which equally turned travel into a personal adventure.[14] Stevenson also demonstrates the same spirit in several passages of his second narrative, *Travels with a Donkey in the Cévennes*, published in 1879.

Compared to touristic forms of travel, these practices were articulated around a different paradigm. While the former relied on a network of fixed places where the traveller could find the comfort and sociability of home recreated, the latter defined itself by the very act of moving. This implied finding food and shelter in villages lying on the way or cooking by the means of a small transportable apparatus and camping in nature. If Stevenson explored this way of living in both his journeys, it is in his second narrative that he manages to capture this attitude to travel through words: 'For my part, I travel not to go anywhere, but to go. I travel for travel's sake. The great affair is to move; to feel the needs and hitches of our life more nearly; to come down off this feather-bed of civilisation, and find the globe granite underfoot and strewn with cutting flints' (Stevenson 2004: 35). The author's conception of travelling crystallizes in the form of an opposition based on Victorian cultural representations. If society provides a protective space allowing people to thrive, this shelter appears to weaken individuals and the expression of nature within them. On the other hand, the journey is presented as an exploration of spaces situated on the outskirts of civilization, where comfort gives way to a rough contact with the world. As the violence of this encounter is understood as archaic, Stevenson borrows images from natural history and imagines a soil scattered with the tools of *man*'s ancestor. Such a representation possesses a strong gendered dimension. It also makes a clear distinction between the author's adventure and his experience of medical travel, which exposed his vulnerability.

Considered from the perspective of travelling on waterways, this definition necessarily excludes the use of steamer or barge, two of the other modes of transport mentioned above as objects of reflection for Stevenson. One way in which these boats especially failed to provide the kind of immediacy with the world he looked for through travel was their reliance on forces other than the traveller's body. As opposed to the image of the tourist-traveller seeking pleasure in comfort, Stevenson's mode of transport instead involved physical activity.

As accounts of adventure journeys like Stevenson's often shift the perspective away from landscape to place the focus on the act of moving forward, they provide detailed descriptions of the operations accomplished by the traveller to make his way on water or land. In the case of canoeing, this implied managing a variety of obstacles, from objects obstructing the way to crossing shallow waters or rapids. The response to these situations required technical skills. MacGregor chose to present them in detail along with other practical information in a separate section placed at the end of his first narrative (1866: 291); but the application of these technical gestures in the action was also particularly interesting for the narrative itself, as it contained an element of drama. At some point, Stevenson narrates facing a difficult section on the Oise, whose power had increased due to heavy rains. The summary of the day's paddling then turns into a list of challenges encountered by the traveller and the response for each case:

> Every two or three hundred yards a tree had fallen across the river and usually involved more than another in its fall. Often there was free water at the end, and we could steer round the leafy promontory and hear the water sucking and bubbling among the twigs. Often, again, when the tree reached from bank to bank, there was room, by lying close, to shoot through underneath, canoe and all. Sometimes it was necessary to get out upon the trunk itself and pull the boats across; and sometimes, where the stream was too impetuous for this, there was nothing for it but to land and 'carry over.' This made a fine series of accidents in the day's career, and kept us aware of ourselves. (Stevenson 1878: 110–11)

As each sentence conveys a new situation for the reader to imagine, Stevenson builds a compelling picture of the variety of experiences involved in canoeing; but overall, the passage provides a sense of the way his mode of travel puts the forces of the body at work. Along with the manoeuvres accomplished by the traveller to guide his canoe, he is also successively shown leaning below a tree or climbing on it or stepping out of the water to carry his boat. All these tasks repeated through a long stretch of the journey are depicted as valued forms of exertion. When Stevenson's health deteriorated, he could be forced to remain immobile for long periods of time. While this lack of activity was the effect of exhaustion, rest was also often the only medicine prescribed by doctors. On the contrary, in his narrative Stevenson follows others before him in presenting canoeing as a healthy activity.

Yet accounts of canoe journeys often record moments when the action goes one step further and includes a notion of danger. It is such an image of his trip that MacGregor presents in the first chapter of his narrative, where he discusses the advantage of the canoe over other boats in providing access to otherwise impracticable parts of rivers. More than an experienced paddler, MacGregor depicts himself as an adventurer, boasting his feats of strength and dexterity in the face of death: 'With one powerful sweep of his paddle he [the canoeist] can instantly turn the canoe, when only a foot distant from fatal destruction' (MacGregor 1866: 2). Such an exaltation of masculine power and mastery shows that canoe adventures reproduced the spirit of the colonial enterprise in the minds of European practitioners and readers.[15] That MacGregor saw himself exploring foreign territory is particularly visible in the narrative of his journey in the Middle East (1869). However, as Stevenson's definition of his travels has shown, British travellers didn't need to go as far to consider themselves stepping into uncivilized territory – although Stevenson adopts a distinct, friendly attitude towards the inhabitants of the countryside.

In *The Unknown River*, Philip Gilbert Hamerton similarly defines his enterprise as an exploratory journey, where an essential part of the pleasure of travelling comes from the act of discovery (Hamerton 1871: 3–5). The notion of danger also emerges in the pages of this account, but the author displays a much less confident attitude towards it. At some point in the journey, Hamerton agrees to carry a passenger, although he is reluctant, knowing his canoe is too small and the man cannot swim: '— Would I take a passenger? He would be delighted to go with me to Toulon if I would give him a berth. — Could he swim? — Swim! not in the least, but he would risk the adventure nevertheless' (ibid.: 49). The scene that follows is the narration of an announced disaster that could have cost the passenger dearly if it were not for his life jacket.

Stevenson's narrative contains its own share of danger, towards which the author shows a mix of boldness and total lack of control, making his journey less the work of a professional adventurer than that of an apprentice traveller in search of sensations. The major accident comes just after the summary of the day's travel quoted above. As he tries to pass below a tree, Stevenson gets stuck, and his boat eventually continues its way down the river without him. In a style that anticipates the author's later adventure novels, the narration captures the tension of that moment when death came looming:

> I had my backboard down in a trice, and aimed for a place where the trunk seemed high enough above the water, and the branches not too thick to let me slip below. When a man has just vowed eternal brotherhood with the universe, he is not in a temper to take great determinations coolly, and this, which might have been a very important determination for me, had not been taken under a happy star. The tree

> caught me about the chest, and while I was yet struggling to make less of myself and get through, the river took the matter out of my hands, and bereaved me of my boat. The Arethusa swung round broadside on, leaned over, ejected so much of me as still remained on board, and thus disencumbered, whipped under the tree, righted, and went merrily away down stream. (Stevenson 1878: 111–12)

The flow of the river and the obstacles that lie on its surface have a direct impact on the experience of the traveller and the content of the narrative. Paddling in a fast current while having to complete technical manoeuvres forces Stevenson to fully concentrate on navigation. As mind and body become oriented towards this single purpose, the aesthetic perception of the landscape disappears, and the narration narrows on action. The rapid succession of decisions immediately followed by execution produces an acceleration of time while the narrative turns into a technical adventure. If such scenes and their aesthetic are typical of canoe narratives, they also represent exceptions in a journey that is most of the time defined by a progression at a regular pace on a relatively peaceful river. As Stevenson resumed this usual rhythm, the contrast between the two experiences appeared to him more clearly. This led him to reflect further on the experience provided by this regular progression.

## Meditations on the Move: The River as a Metaphorical Locus

In the major part of the journey, Stevenson's quiet and consistent pace on the river left him free to look at the landscape around him, which found its way into the narrative in the form of descriptions. Instead of entirely returning to this habit, as paddling goes back to being merely repetitive and deprived of complexity, the author contemplates his own mental state. In what corresponds to a new phenomenological examination of river navigation, this time centred on the experience of canoeing, he considers the hypnotizing effect of his mechanical dipping of the paddle in the water on alternate sides:

> This frame of mind was the great exploit of our voyage, take it all in all. It was the farthest piece of travel accomplished. Indeed, it lies so far from beaten paths of language, that I despair of getting the reader into sympathy with the smiling, complacent idiocy of my condition; when ideas came and went like motes in a sunbeam; when trees and church spires along the bank surged up, from time to time into my notice, like solid objects through a rolling cloudland; when the rhythmical swish of boat and paddle in the water became a cradle-song to lull my thoughts asleep; when a piece of mud on the deck was sometimes

> an intolerable eyesore, and sometimes quite a companion. (Stevenson 1878: 200)

In this passage, Stevenson depicts a moment of climax in his canoe journey. After that, the few remaining chapters are only dedicated to the last places visited, but the progression by canoe no longer stands at the forefront of the narrative. Here the body almost entirely disappears from the picture as the mind experiences a form of levity, leaving the traveller barely aware of the world around him. Interestingly, the act of moving away from the beaten tracks of common tourist experience is translated into aesthetic terms, as the difficulty of capturing this state of mind through words is compared to an exploration of the boundaries of language. Although the author presents this unique moment as an epiphany, his self-reflexive meditation upon his experience underlines another dimension of his journey.

When Stevenson defined his narrative as 'the history of a man's life during some weeks of travel' in his letter to Colvin (Booth and Mehew 1994: 231–32), he did not only understand it as a mere factual compilation of a series of accidents and events. His vision also included the internal life of the traveller and its evolution throughout the journey. This other landscape takes the shape of the various thoughts and comments that the author shares in response to the elements that present themselves to his senses. In the passage above, the very special state in which Stevenson finds himself does not allow him to form any complex chains of thought, but in the rest of the narrative his regular progression leaves him plenty of time to meditate. Carried by the gentle pace of the river, his reflections grow from his immediate experience until the discussion encompasses much broader aspects of Victorian society or human life.

That is where the reference to Heine's *Reisebilder*, or *Travel Pictures* (2008), as a major source of inspiration for Stevenson, comes to make sense. This work contains the narratives of journeys completed at different points in the life of the romantic poet; but more than coming from the places visited, the force of Heine's work arises from his witty and often humoristic thoughts, which set the spirit of the whole book. These comments and broader meditations find their place in the narrative of the journey, but behind them stands another genre that was part of Heine's literary practice: the essay. Stevenson was also a brilliant essayist. This genre had formed the core of his literary endeavours and continued to do so even after he illustrated himself as a writer of fiction. It thus comes as no surprise that, like Heine, this vein should appear in his travel writing.[16]

*An Inland Voyage* includes developments on aesthetic or ethical matters, sometimes in a lyrical tone. Compared to comments on navigation or descriptions of remarkable encounters and events, such discursive passages are less central to narratives of river journeys as a genre. Many travellers, like Edmund Harvey, rather gave importance to their itinerary and progression, and defined their work more narrowly as 'a simple and unvarnished account of [a] cruise' that 'makes no attempt at flowing language, or even a description of the beauties of scenery' (Harvey 1854: 3). Such a straightforward tone could be the strategy adopted by some, like MacGregor, to fashion themselves as adventurers. If Stevenson occasionally enjoys embodying that figure in his own way, most of the time he prefers adopting the romantic posture of the poet traveller or contemplator. Like in his essays, his meditations cover a great variety of topics, but in the context of the journey they emerge in the form of a succession, as he makes his way forward. Among all these reflections, some directly spring from the movement of his canoe. Following the flow of the river, Stevenson's thoughts crystallize on concerns about the course of life, time and the rhythm of the society he lives in.

If these themes emerge in response to his immediate environment, they also connect to a broader Western tradition that the Victorians contributed to shaping through constant reference to ancient texts in their works. In this tradition, the river figures as a symbolic locus. That the author carried representations borrowed from the classics appears through the direct reference to Pan in the narrative. The mythological figure can be encountered in the landscape of the Oise, where it is depicted as having been heard by generations of reeds: '*Pan* once played upon their forefathers; and so, by the hands of his river, he still plays upon these later generations down all the valley of the Oise; and plays the same air, both sweet and shrill, to tell us of the beauty and the terror of the world' (Stevenson 1878: 104). The god also features on the frontispiece of the book, lying by the river with the two canoeists in the background. A few months after the publication of his account, Stevenson even wrote a separate essay, 'Pan's Pipes', which was later included in his collection *Virginibus Puerisque* (Stevenson 2018: 106). However, beyond this figure, it is in the motifs associated with the river that the author conveys classical images in his meditations.

In a passage in which he observes the neighbourhood formed by the still line of barges on the canal, Stevenson points out the fake impression of permanence conveyed by this picture, reminding his readers that these itinerant houses may well break at any time and go in separate directions. Then, contemplating the life provided by barges, Stevenson finds in it an illustration of the transience of things, a topic that bears Heraclitean resonance. His meditation concludes on an image full of pathos, through which he underlines the tragedy of this life for the youngest inhabitants on board: 'The children who played together to-day by the Sambre and Oise Canal, each at his own father's threshold, when and where might they next meet?' (Stevenson 1878: 93). As Stevenson considers the uncertainty of a future reunion between the neighbours, the river turns into a spatial and temporal symbol of the uncertain course of human life.

If this understanding of the river as a metaphorical locus escapes the concerns of more adventurous narratives, Philip Hamerton similarly plays on the same images in his account. Chapter VIII of *The Unknown River* opens on a meditation that concludes on an aphorism tying several aspects of the author's experience together: 'the most interesting life to live, like the best river to explore, is one in which the course is not known in detail beforehand, but constantly calls for the exercise of skill and judgement and is even to some degree affected also by pure hazard' (Hamerton 1871: 43). As an artist and critic, Hamerton had a similar profile to Stevenson's. The two not only targeted an audience that was looking for the spirit of adventure in travel narratives; they also wrote for an elite readership for whom such philosophical reflections determined the value of a work. Actualizing classical representations of the river by anchoring them in the contemporary experience of the traveller was one way of providing that intellectual perspective. However, behind the general dimension of Stevenson's meditations lies a critical comment about his society, especially its rhythm.

Like John MacGregor, in his account Stevenson records instances of popular enthusiasm for the travellers and their canoes in the villages they visit. This exaltation occasionally shows as the crowd gathers by the river in the morning to see the two figures leave. Alongside these lively scenes, more quiet departures in the early hours of the morning enable the author to plunge into his meditative mood. On one such occasion, he comes to think about the villagers as people he himself shall never see again. He then reflects on the sensation provided by these repeated departures and on the articulation of his journey within the normal course of life: 'It gives the traveller a jog, reminds him that he is not a traveller everywhere, and that his journey is no more than a siesta by the way on the real march of life' (Stevenson 1878: 207). Like the river, life appears as a continuous movement forward, in which travelling introduces a change of pace or a temporary break.

This notion of rhythm also proves central in Stevenson's meditations on barges. In the continuation of his previous reflections, the author recalls that he and his companion started imagining spending their old days on one of these boats: 'It was to be the most leisurely of progresses, now on a swift river at the tail of a steam-boat, now waiting horses for days together on some inconsiderable junction' (Stevenson 1878: 93). In this ideal picture, pleasure comes from the change of pace applied to life's march forward. Alongside moments of fast progress, Stevenson especially underlines the value of letting time go to waste.[17] This idea went completely against the principles of Victorian society, which had been dominated by the spirit of industrial capitalism and was witnessing the early developments of global finance at the time when the author was writing.

Stevenson had experienced that reality the hard way in the years before his journey. Initially full of romantic ideals about literary creation, he had discovered the pressure of having to write for a market. In that context, he later came to consider literature just another 'business' in his society (Booth and Mehew 1994: 290). Although he had no choice but to abandon some of his ideals, the apprentice author took a stance against the spirit of his time in his writings and asserted the value of slowly wandering on the path of life. In *An Inland Voyage*, this attitude is crystallized both in his itinerary and mode of transport. If at one point the two travellers favour the train for its efficient linearity, during the rest of the journey they follow the serpentine course of the river and let themselves be carried by the varying rhythm of the water-flow.

That this attitude to travel was connected to a critical view of Victorian society particularly appears in a comment as the author depicts himself enjoying a break by the river: 'At that hour, stockbrokers were shouting in Paris Bourse for two or three per cent; but we minded them as little as the sliding stream, and sacrificed a hecatomb of minutes to the gods of tobacco and digestion' (Stevenson 1878: 148). This passage echoes two essays that Stevenson had written the year before the publication of *An Inland Voyage*. In 'The Paris Bourse' (Stevenson 2021: 360), the author embodies the figure of the idler considering the atmosphere inside the financial institution from a safe critical distance, from which he perceives the place as something close to Dante's *Inferno*. In the same period, Stevenson captured his attitude to life in his essay 'An Apology for Idlers' (Stevenson 2018: 45). There he argues that idling, far from being an empty activity, can provide access to a knowledge that escapes the busy and well-respected fellow.[18] If the author could defend the value of idling, it is because he was no stranger to hard work. He also found it more difficult to play the aesthete when pressed by necessity, in which case he could still count on the financial help provided by his parents, who had thrived thanks to the family's engineering business. That said, by looking for knowledge outside the sphere of work or established institutions to locate it in experience, Stevenson manages to complete the theoretical frame that sustains his travels. The narrative of his canoe journey thus offers more than just an adventure in foreign territory, it associates a particular spirit with an emerging practice and defines a special attitude towards life during a momentary break from business.

## Conclusion: Leisure and the Pursuit of Knowledge

Stevenson's enterprise finds its place in the landscape of Europe in the second half of the nineteenth century. This period was defined by the increase of touristic mobility in Britain and on the continent. If this dynamic was mainly sustained by the expansion of the railways, waterways came to be fully integrated into the growing network of pleasure travel. However, distinct from mainstream forms of tourism, the author's journey partook in the development of a new type of recreational practices. Reserved for more audacious individuals, outdoor adventure associated travelling for pleasure with physical effort and immediate contact with nature. Depicting the journey as an expedition, this form of travel brought home the masculine spirit of conquest attached to the colonial enterprise. Like a few others, Stevenson carried out this journey by canoe, a boat that had recently become integrated into Europe as a tool for recreation.

*An Inland Voyage* represents a contribution to the literature that emerged from such adventures on waterways. While the author joined others in infusing parts of his narrative with the spirit of romance, he also developed a personal attitude to his journey, from which comes the specific tone of his account. Before being an adventurer Stevenson was a contemplator, and it is in that frame that he associated travelling with slow and sinuous progress. He considered this rhythm not only part of a break from the usual pace of life, but a condition for the journey to become a place of learning. As such, in his account the author redefines the boundaries between work and leisure, and renews the articulation of the latter with the classic notion of *otium*, that is, with the pursuit of knowledge rather than mere entertainment.

This conception of the journey acknowledges the transformative power of travelling on the individual. It also reminds us that Stevenson's adventure was the work of someone trying to articulate his experience with the one provided by modern means of transportation. While steam-powered engines were seen as an important step in the unstoppable march of progress, announcing the special place held by speed in the twentieth-century experience of modernity, in his first two journeys Stevenson takes a stance against this vision by using forms of mobility that he also perceived as archaic. As such, through his travels, the former apprentice engineer takes his distance from the general spirit of celebration of futurity to contemplate the disappearance of a world belonging to the past while some remnants of it, like his canoe, come to bear new relevance in the present. This superimposition of several temporalities in the narrative continues to speak to contemporary readers. At a time when recreational mobilities are now entangled with the broader issue of the climatic and environmental crisis, slow modes of travel and the particular experience that they provide continue to acquire new meaning.

**Kévin Cristin** teaches at the Department of Anglophone Studies at Aix-Marseille University, France, where he completed his PhD in 2024. His research focuses on the intersections between literature, science and technology in the nineteenth century, especially in the works of Robert Louis Stevenson, the popular writer and member of a dynasty of engineers. His dissertation examines how the author's experience as an apprentice engineer shaped significant features of his literary works. Cristin's investigation follows the itinerary of Stevenson across disciplines, from his early training in engineering to his anthropological work in the colonial context of the Pacific.

## Notes

1. Evans (1981) provides a detailed overview of the technical reasons that led to this shift in Britain.
2. For first-hand accounts of the practice, see Woodgate (1888, 1889).
3. For lack of any better or scholarly edition, I will use the original account published by Stevenson.
4. This experience was described in 'Ordered South', in Stevenson (2018).
5. In the letters, I will only be interested in a few passages where Stevenson mentions his journey and book project. On travel writing in epistolary form per se, see Bannet (2019).
6. Erickson and Krotz discuss the colonial history of the canoe and its early recreational use in nineteenth-century North America as part of a larger study on the continuing strong cultural and political dimension of canoeing today for Indigenous communities in the region.
7. In *The Last of the Mohicans* (2008), the author describes an action scene involving a canoe race between the protagonists and the natives.
8. On the connection between the works of Cooper and Scott, see Pilote (2022).
9. On this type of literature and its proximity to travel accounts in the nineteenth century, see Pettinger (2019).
10. MacGregoralso inspired other travellers, like James Henry Hamilton Gordon or George Heaviside, whose narratives of canoe journeys in Europe, respectively published in 1868 and 1871, are mentioned by Vine (1983: 122, 145) but do not seem to have reached such a wide audience at the time of their publication.
11. See 'Walking Tours', later included in Stevenson (2018).
12. Later, in 1882, Stevenson titled an essay on his fictional genre of predilection 'A Gossip on Romance'.
13. The book was published anonymously, the author being referred to as 'the captain.' It was current practice for writers of accounts of river journeys to use pseudonyms. In *An Inland Voyage* Stevenson and Simpson are themselves named after their canoes, the *Arethusa* and the *Cigarette*.
14. For a recent account of mountaineering considering this dimension, see Schaumann (2020).
15. This spirit was also part of a longer tradition. Sarah Goldsmith (2020) has shown that it already represented a central aspect of the experience of the Grand Tour in the eighteenth century.
16. *Essays I: Virginibus Puerisque and Other Papers* (Stevenson 2018: xxv–lxviii) provides a comprehensive list of Stevenson's essays and a rich discussion about the author's contribution to the genre.
17. For a larger reflection on time in Stevenson's travel narratives, see Davies (2009: 73).
18. On the association between travel and idling beyond Stevenson, see Liedke (2018), which includes a discussion of Jerome K. Jerome's humorous narrative, *Three Men in a Boat*, published in 1889.

## References

Anderson, Susan C., and Bruce H. Tabb (eds). 2002. *Water, Leisure and Culture: European Historical Perspectives*. Oxford: Bloomsbury Academic.

Baden-Powell, Warington. 1871. *Canoe Travelling: Log of a Cruise on the Baltic, and Practical Hints on Building and Fitting Canoes*. Smith: Elder.

Baldwin, Mark. 1984. *Canal Books: A Guide to the Literature of the Waterways*. London: M & M Baldwin.

Baldwin, Mark, and Anthony Burton (eds). 1984. *Canals: A New Look*. Chichester: Phillimore.

Bannet, Eve Tavor. 2019. 'Letters', in Alasdair Pettinger and Tim Youngs (eds), *The Routledge Research Companion to Travel Writing*. New York: Routledge, pp. 115–27.

Barton, Susan, and Allan Brodie (eds). 2021. *Travel and Tourism in Britain, 1700–1914*. 4 vols. London: Routledge.

Booth, Bradford Allen, and Ernest Mehew (eds). 1994. *The Letters of Robert Louis Stevenson*. New Haven, CT: Yale University Press.

Buckton, Oliver S. 2007. *Cruising with Robert Louis Stevenson: Travel, Narrative, and the Colonial Body*. Athens: Ohio University Press.

Burroughs, Robert. 2019. 'Travel Writing and Rivers', in Nandini Das and Tim Youngs (eds), *The Cambridge History of Travel Writing*. Cambridge: Cambridge University Press, 330–44.

Buzard, James. 1993. *The Beaten Track: European Tourism, Literature, and the Ways to Culture, 1800–1918*. Oxford: Clarendon Press.

Cooper, James Fenimore. 2008 (1826). *The Last of the Mohicans*. Oxford: Oxford University Press.

Cusack, Janet. 2000. 'Nineteenth-Century Cruising Yachtsmen in the Mediterranean', *Journal of Mediterranean Studies* 10(1): 47–75.

Davies, Laurence. 2009. 'The Time of His Time: *Travels with a Donkey* and *An Inland Voyage*', in Richard Ambrosini and Richard Dury (eds), *European Stevenson*. Newcastle-upon-Tyne: Cambridge Scholars, pp. 73–89.

Erickson, Bruce, and Sarah Wylie Krotz (eds). 2021. *The Politics of the Canoe*. Winnipeg: University of Manitoba Press.

Evans, Francis T. 1981. 'Roads, Railways, and Canals: Technical Choices in 19th-Century Britain', *Technology and Culture* 22(1): 1–34.

Fisher, Stephen. 1997. *Recreation and the Sea*. Exeter: University of Exeter Press.

Goldsmith, Sarah. 2020. *Masculinity and Danger on the Eighteenth-Century Grand Tour*. London: University of London Press.

Hamerton, Philip Gilbert. 1871. *The Unknown River*. London: Seeley.

Harvey, Edmund George. 1854. *Our Cruise in the Undine: The Journal of an English Pair-Oar Expedition through France, Baden, Rhenish Bavaria, Prussia, and Belgium*. London: J. W. Parker and Son.

Heine, Heinrich. 2008 (1826). *Travel Pictures*, trans Peter Wortsman. Brooklyn: Archipelago.

Liedke, Heidi. 2018. *The Experience of Idling in Victorian Travel Texts, 1850–1901*. Cham: Palgrave Macmillan.

MacGregor, John. 1866. *A Thousand Miles in the Rob Roy Canoe on Rivers and Lakes of Europe*. London: Sampson Low, Son, and Marston.

———. 1867. *The Rob Roy on the Baltic: A Canoe Cruise through Norway, Sweden, Denmark, Sleswig, Holstein, the North Sea, and the Baltic*. London: Sampson Low, Son, and Marston.

———. 1869. *The Rob Roy on the Jordan, Nile, Red Sea & Gennesareth, &c.: A Canoe Cruise in Palestine and Egypt, and the Waters of Damascus*. London: J. Murray.

Mansfield, Robert Blachford. 1852. *The Log of the Water Lily-Four-Oared Thames Gig-during a Rowing Excursion on the Rhine, and Other Streams of Germany*. London: John W. Parker & Son.

———. 1853. *The Water Lily on the Danube: Being a Brief Account of the Perils of a Pair-Oar during a Voyage from Lambeth to Pesth*. London: John W. Parker & Son.

Moens, William John Charles. 1876. *Through France and Belgium by River and Canal, in the Steam Yacht 'Ytene'*. London: Hurst & Blackett.

Molloy, James L. 1874. *Our Autumn Holiday on French Rivers*. London: Bradbury, Agnew & Co.

Pettinger, Alasdair. 2019. 'Guidance and Advice', in Alasdair Pettinger and Tim Youngs (eds), *The Routledge Research Companion to Travel Writing*. New York: Routledge, pp. 139–52.

Pilote, Pauline. 2022. *Wizards of the West: Walter Scott et le roman historique américain*. Rennes: Presses Universitaires de Rennes.

Schaumann, Caroline. 2020. *Peak Pursuits: The Emergence of Mountaineering in the Nineteenth Century*. New Haven, CT: Yale University Press.

Simpson, Evelyn Blantyre. 1912. *The Robert Louis Stevenson Originals*. London: T. N. Foulis.

Stevenson, Robert Louis. 1878. *An Inland Voyage*. London: Kegan Paul & Co.

———. 2004 (1879). *Travels with a Donkey in the Cévennes and the Amateur Emigrant*. London: Penguin Classics.

———. 2018 (1881). *Essays I: Virginibus Puerisque and Other Papers*, ed. Robert-Louis Abrahamson. Edinburgh: Edinburgh University Press.

———. 2021. *The Complete Personal Essays of Robert Louis Stevenson*, ed. Trenton B. Olsen. Abingdon: Routledge.

Vine, Paul A. L. 1983. *Pleasure Boating in the Victorian Era: An Anthology of Some of the More Enterprising Voyages Made in Pleasure Boats on Inland Waterways during the Nineteenth Century*. Chichester: Phillimore.

Walton, John K. 1983. *The English Seaside Resort: A Social History, 1750–1914*. Leicester: Leicester University Press.

———. (ed.). 2016. *Mineral Springs Resorts in Global Perspective: Spa Histories*. London: Routledge.

Wenham, Simon. 2014. *Pleasure Boating on the Thames: A History of Salter Bros, 1858–Present Day*. Stroud: History Press.

———. 2020. *Hobbs of Henley: A History*. Stroud: Amberley.

Wigglesworth, Neil. 2013. *The Social History of English Rowing*. London: Routledge.

Woodgate, Walter Bradford. 1888. *Boating*. London: Longmans, Green and Co.

———. 1889. *Rowing and Sculling*. London: George Bell & Sons.

# CHAPTER 3

# F. Scott and Zelda Fitzgerald's Literary and Visual European Journeys

## *A 'Carnival by the Sea', and Many 'Quick Odysseys'*

Elisabeth Bouzonviller

### Introduction

I'd like to live and die on the French Riviera.
—F. Scott Fitzgerald, Letter to Maxwell Perkins, 20 February 1926

F. Scott and Zelda Fitzgerald are not widely remembered as travel writers since only a couple of their short memoirs published in various magazines during the 1920s and 1930s can be considered travelogues. However, as de Certeau claims, 'all narratives are travel narratives – a practice of space' (de Certeau 1990: 71).[1] Moreover, the Fitzgeralds were relentless travellers and their fiction, correspondence, self-narratives and Zelda's paintings reflect their ceaseless restlessness and desire to refer to it artistically. Their scrapbooks[2] are also evidence of their nomadic existence. In her introduction to *The Romantic Egoists: A Pictorial Autobiography from the Scrapbooks and Albums of F. Scott and Zelda Fitzgerald*, their daughter Scottie explains that she inherited 'seven scrapbooks and five photograph albums' from her parents: 'it is extraordinary how meticulous my father was about making and keeping personal records. The first scrapbook takes him from birth into the Army, the second takes my mother from birth through the first few years of marriage, and the remaining five contain reviews, interviews, letters, pictures, and memorabilia of all varieties' (Bruccoli 2003: ix). Thanks to all sorts of visual and written elements, these scrapbooks record their constant national and international travelling; they are a recollection of their married and then family life on the move.

The Fitzgeralds' lack of geographical anchoring is underscored by the fact that they never owned a house or apartment of their own throughout their lives despite times when their finances could have made it possible. On the other hand, they owned several successive cars, in the US and Europe, as testified by photographs, narratives and various official documents.[3] From the start, theirs is a story of movements and journeys.

Born in St Paul, Minnesota, in 1896, Scott met Zelda, one of the most popular Southern Belles from Montgomery, Alabama,[4] in 1918, during his military training. They eventually married in New York, away from their families, on 3 April 1920, and by 3 May 1921, they were on their way to Europe for the first time. Zelda had known since February that she was pregnant and they wanted to travel before her pregnancy was too advanced. This was the first of their four transatlantic trips, which in the end led them to spend almost six years in Europe, mainly in France, but also in England, Switzerland and Italy.[5]

Relying on their correspondence, but also on Scott's self-narratives *The Cruise of the Rolling Junk*, 'How to Live on Practically Nothing a Year' and 'The High Cost of Macaroni' and Zelda's '"Show Mr. and Mrs. F. to Number—"' and 'Auction—Model 1934', together with their 1930s novels – *Tender Is the Night* (1934) and *Save Me the Waltz* (1932) – and a few short stories, much inspired by their transatlantic journeys in terms of settings and atmosphere, I intend to show how they pictured a restless transmedial modernity and a 'shrinking' globe,[6] as predicted by Henry James in 1898. Moreover, their scrapbooks offer precious additional material to assess the rendering of their travel experiences. Barbara Korte defines travelogues on the basis of two conditions: 'the journey recorded actually took place', and that journey 'is presented by the traveller him or herself' (Korte 2000: 1), which corresponds to the self-narratives and pictures in question here. Always keen on staging themselves and displaying a certain public image, the Fitzgeralds advertised themselves on the move, and transatlantic trips were certainly part of the process. This was performed thanks to their writings, whether fictional or autobiographical, but also to the numerous visual documents they meticulously kept, organized and sometimes published in the press, which constitute an unconventional travel literature corpus in keeping with an age that saw the rise of advertising techniques and of the Hollywood star system that kept audiences informed of their idols' restless lives.

In *Tender Is the Night*, the narrator describes the landscape from the Glion funicular in Switzerland and concludes: 'On the centre of the lake, cooled by the piercing current of the Rhône, lay the true centre of the Western World' (Fitzgerald 2018: 169). These words seem to imply that Europe embodies a 'centre' and, therefore, sojourns for Americans there would become a geographical decentring, replacing the travellers at the very centre of things. However, in fiction as in life, these expatriates fail to perceive it this way as they never really integrate within European society nor display great interest in European people. Indeed, despite their regular European trips and stays, like their characters, the Fitzgeralds remained fleeting visitors who lacked integration, as shown by their American-dominated surroundings even while abroad. As explained by Fréderic Regard, explorers' narratives stir reflection 'about the varied processes of "othering" the not-Self, but also about self-fashioning' (Regard 2009: 5). Keeping this pattern in mind, even if the Fitzgeralds were not explorers but rather worldly travellers, it can be asserted that their decentring did not imply so much an approach to otherness as a 'self-fashioning', which, by the way, often took place at the expense of the locals. In fact, transatlantic sojourns provided them with a vantage point to ponder their Americanness and inner selves. Eventually, for them, Europe, and France in particular, was mainly a fruitful location that sharpened their artistic skills and shaped them as modernist artists.

## A Family on the Move

As James Mellow explains, the Fitzgeralds were in the spotlight of American artistic life and never forgot to feed and polish their fictional selves: 'The Fitzgeralds had always been public property, adepts at self-advertisement. . . . the ultimate fascination . . . was what masters of invention they became, creating new versions of themselves, putting themselves into their stories, acting out their stories in real life' (Mellow 1985: xvii, xx). Travelling was part of that dramatization of the self. Long before Instagram influencers intent on showing themselves within exciting new environments throughout the globe, the Fitzgeralds staged themselves on the move and kept the public informed of their fashionable whereabouts through texts and pictures, be it thanks to the journalists who reported their activities or to their own publications in the press. Crossing the ocean, driving, taking trains and discovering new cultures were signs of modernity. The Fitzgeralds' geographic dynamism even heralded what Augé calls the 'supermodernity' of the twentieth century (Augé 1995: 93), which is characterized by 'spatial overabundance' and the advent of 'non-places' (ibid.: 32, 75–115).

This urge to travel spread among artists and intellectuals of the 1920s, especially in the United States, which was going through a very paradoxical period involving both great modernity and backward fundamentalism. American critic and essayist Harold Stearns, who left New York in 1921 for Europe, wrote: 'A young man had no future in this country of hypocrisy and repression. He should take ship for Europe, where people know how to live' (Moore 2008: 238). During the post-First World War decade, many would seek novelty and freedom elsewhere, especially in France, including Ernest Hemingway, John

Dos Passos, Ezra Pound, T.S. Eliot, Edmund Wilson, Archibald McLeish and Cole Porter. Some had preceded them, like Nathalie Barney, Edith Wharton, Gertrude Stein and Sylvia Beach, and were making Paris a place for artists to meet and create. In 1922, Beach became the eventual publisher of *Ulysses* by James Joyce, a regular patron of her English-language bookstore Shakespeare and Company, after the work faced censorship elsewhere.

The intellectual vibrancy of Paris as a meeting point for English-speaking modernists can be encapsulated in the Franco-American literary agency that William and Jenny Bradley developed during the 1920s on Saint-Louis Island. Writers travelled, settled for a while in France and sought to be published or to be put in touch with literary figures, which Jenny Bradley did, for example, with Fitzgerald who wished to be introduced to Anatole France (Cossu-Beaumont 2023: 64). Studying this literary agency, Laurence Cossu-Beaumont concludes that 'All those letters sent by transatlantic liners, and later by air, and all those books read on either side of the ocean, form a conversation that continues into the 21st century' (ibid.: 28–29). Focusing on the agency's transatlantic activities, she confirms Pascale Casanova's idea of 'Paris as "Greenwich meridian" of literature' (ibid.: 27–28). In an age that was caught in an ever-increasing acceleration, as predicted by Henry Adams,[7] writers travelled, wrote abroad, and while they were on the move, their books required publication and translation to travel themselves among a wider readership.

Korte explains that 'travel writing is defined by the interaction of the human subject with the world' (Korte 2003: 5). After an 'early success' (Fitzgerald 1965: 57) on the fiction market with his first novel, *This Side of Paradise*, in 1920, Fitzgerald soon learned how to combine his writer's talent with his relentless nomadism and he started to publish self-narratives based on his and his wife's travelling adventures. This was typical of the spirit of the age, as mentioned by Paul Fussell, who contrasts travelling and tourism, and contends that the 1920s and 1930s were 'the final age of travel' as travelogues by successful British authors flourished then (Fussell 1980: vii). According to Fussell, the great age of travelling born out of the confining years of the First World War came to an end with the Second World War due to mass tourism, travelling restrictions and jet planes: 'Clearly not a world in which either travel or the travel book could flourish. The going was good for only twenty years, and after the war all that remained was jet tourism among the ruins, resulting in phenomena like the appalling pollution of the Mediterranean and the Aegean' (ibid.: 226).[8] While James Buzard recalls that, according to the *Oxford English Dictionary*, the term 'tourist' was synonymous for 'traveller' in the late eighteenth century (Buzard 1993: 1), he tries, like Fussell before him, to analyse in what ways the two terms have eventually acquired contrasted meanings. Buzard explains: 'The traveller exhibits boldness and gritty endurance under all conditions (being true to the etymology of "travel" in the word "travail"); the tourist is the cautious pampered unit of a leisure industry' (ibid.: 2). Beforehand, Fussell had nostalgically defined the evolution from exploring, then travelling, towards tourism, thus sketching a kind of 'elegy' of literary travelling: 'I am assuming that travel is now impossible and that tourism is all we have left' (Fussell 1980: 227, 41). Reviewing Fussell's *Abroad*, Samuel Hynes concludes: 'for Fussell, the tourist is only one symptom of the general decline and fall of civility in our time' (Hynes 1982: 368).[9] Transgressing those definitions, the Fitzgeralds were travellers at home, then abroad; they were tourists at times, eager to discover and photograph European sites advertised by the Baedecker Guides, but, most all, they were voluntary expatriates.

In July 1920, before the first trip to Europe, the Fitzgeralds went on a 1,200-mile road trip from Westport, Connecticut to Montgomery, Alabama. The memoir of this car trip was written two years after it took place and published in three instalments in 1924 in the magazine *Motor* and eventually as a single volume, *The Cruise of the Rolling Junk*, in 1976. It was therefore a retrospective narrative devoid of the spontaneity of the moment that tried to reconstruct their earlier travelling spirit and adventures, hence the lack of accuracy and the diverging memories between this travelling account and Zelda's evocation of the trip, which was written even later (Z. Fitzgerald 1992b: 419). Originally, the text was illustrated with gag pictures of the travellers themselves and spread over several pages as the magazine also included many advertisements related to car driving. The Fitzgeralds' pictures displayed their comic adventures and proclaimed them actors of a new age in which cars embodied fashionable modernity. The three essays and pictures form a kind of journalistic travel piece that relies on geographical observations, the burlesque mishaps of the travellers on their way to the South and the pitiless descriptions of the people encountered. *The Cruise of the Rolling Junk* also shaped Fitzgerald's humorous tone and style for his following essays during his European trips.

In 1921, the couple embarked on their first transatlantic trip, but it was a disappointment. Admired and celebrated in the US as the glamorous embodiment of the Jazz Age, they met with less enthusiasm in Europe. Having left on board the *Aquitania* on 3 May, they arrived in Southampton on 10 May, and reached Paris on 17 May. In a letter to Shane Leslie dated 24 May, Fitzgerald expressed his implacable judgement: 'France is a bore and a disappointment chiefly, I imagine, because we know no one here. Italy on Wednesday' (Turnbull 1964: 379). They had, indeed, obtained visas to go to Italy, which they reached on board the Simplon Orient-Express train through the Swiss border, as shown by the visa document that Zelda pasted in her scrapbook. On 26 May, they were in Venice, on 3 June in Florence and in Rome on 8 June. Their visa document shows that they went through the French border at Modane in Savoy on their way back to Paris on 22 June. In Zelda's scrapbook, a few tourists' pictures with humorous comments recall this quick discovery of Europe. Nevertheless,

Italy seems to have rated no better than France in their eyes since, in July, Fitzgerald wrote to his friend Edmund Wilson, and ranted against the European countries he had visited in a racist and violent paragraph that ranks only with his later praises in terms of excess: 'God damn the continent of Europe. It is of merely antiquarian interest. Rome is only a few years behind Tyre and Babylon. ... France made me sick' (ibid.: 326).

However, in 1924, they were back, as a family this time, since their daughter Scottie was travelling with them. That same year, Fitzgerald published two self-narratives in the *Saturday Evening Post*: 'How to Live on $36,000 a Year' on 5 April, and 'How to Live on Practically Nothing a Year' on 20 September. These essays tackle the topic of money and travelling, and how they decided to move to France in early May of that same year. France was, indeed, a favourable economic solution given the rate of the dollar against the franc at the time; it was a way to 'economize', as is repeated in the latest essay (Fitzgerald 1957b: 101, 105, 114). It was also a second opportunity to experience France, which Fitzgerald would now describe in a completely new way. In 1926, he wrote to his editor Maxwell Perkins: 'I'd like to live and die on the French Riviera' (Turnbull 1964: 199). And in a 1927 interview he declared: 'The best of America drifts to Paris. The American in Paris is the best American. It is more fun for an intelligent person to live in an intelligent country. France has the only two things toward which we drift as we grow older – intelligence and good manners' (Bruccoli 1991: 299).

In *Paris France*, Gertrude Stein argues that 'Paris was the place to be' (Stein 2003: 13). However, in 1924, the Fitzgeralds, who would meet her through Hemingway later on in 1925, reached the capital, looked for a nanny for their daughter and quickly left for the French Riviera. In 'How to Live on Practically Nothing a Year', the narrator mentions their escape from a Paris crowded with Americans and celebrates 'the hot, sweet South of France': 'I think now that we did well to get away from Paris' (Fitzgerald 1957b: 104). There follow two enthusiastic paragraphs about the charms of the Mediterranean and its resorts, with insistence on the impressive, visual, even sometimes cinematic quality of the landscape:

> It is a blue sea; or rather it is too blue for that hackneyed phrase which has described every muddy pool from pole to pole. It is the fairy blue of Maxfield Parrish's pictures; blue like blue books, blue oil, blue eyes, and in the shadow of the mountains a green belt of land runs along the coast for a hundred miles and makes a playground for the world. The Riviera! (Fitzgerald 1957b: 104)

The repetition of the word 'blue', the evocation of the neoclassical pictorial expert and the exclamative form all emphasize the viewer's excitement and delight. Following one of the travel writing norms, his gaze aestheticizes the landscape. In his correspondence, he was no less enthusiastic and hyperbolic, as when he wrote, in May 1924, to Thomas Boyd: 'This is the lovliest [*sic*] piece of earth I've ever seen without excepting Oxford or Venice or Princeton or anywhere. Zelda and I are sitting in the café l'Universe writing letters . . . and the moon is an absolutely *au fait* Mediteraenean [*sic*] moon with a blurred silver linnen cap [*sic*].'[10] In the same vein, in June 1924, he wrote to Carl Van Vechten: 'We are living here in a sort of idyllic state among everything lovely imaginable in the way of Mediterranean delights' (Turnbull 1964: 477), and to Maxwell Perkins: 'We are idyllically settled here' (ibid.: 164).

Ten years later, he would rely on this 'hot sweet South' (Fitzgerald 2018: 40) to open his fourth novel, *Tender Is the Night*, with an equally picturesque description, inspired by Cap d'Antibes and Juan-les-Pins:

> On the pleasant shore of the French Riviera, about half way between Marseilles and the Italian border, stands a large, proud, rose-colored hotel. . . . In the early morning the distant image of Cannes, the pink and cream of old fortifications, the purple Alp that bounded Italy, were cast across the water and lay quivering in the ripples and rings sent up by sea-plants through the clear shallows. (Fitzgerald 2018: 5)

Impressed by an environment that would haunt him long after he had left France forever, the novelist transformed himself into a painter relying on all the colours available on his palette to express the beauty of a place he wanted to celebrate as visitor, temporary resident and novelist. However, it would remain a kind of exotic backdrop enabling him to focus on American behaviours and issues, as his narratives and essays betray a lack of real interest in cultural otherness and of effort at personal integration.

## Americans Abroad

As Emmanuelle Peraldo argues, 'the writing of space is never purely aesthetic, but always ideological or political' (Peraldo 2016: 11). Writing about their European journeys, the Fitzgeralds definitely remained American and asserted their authority as such. Pierre-Étienne Vandamme defines decentring as different from empathy or altruism, but rather as an interest in otherness and 'the ability to distance oneself from one's bearings, one's convictions, one's vision of the world, in order to meet others' (Vandamme 2017: 168).[11] Nothing of the kind happens in *Tender Is the Night* nor in the Fitzgeralds' life when they stayed in France. Vandamme argues that 'decentring marks an interest in otherness and an attempt to avoid examining others from an *ego*- or *ethno*-centric perspective' (ibid.: 170).[12] Despite the pleasure of their French stays and enduring memories of them, the Fitzgeralds remained Americans abroad, often displaying an attitude of dominance, which Scott openly acknowledges in 'How to Live on Practically Nothing a Year' when he writes 'I felt very American and superior' (Fitzgerald 1957b: 106). To sum up, they enjoyed the place but ignored, even sometimes scorned, the inhabitants. However, at the end of this essay, after months on the Riviera, the narrator contradicts himself and concludes: 'We've become absolutely French' (ibid.: 113). This, of course, was an illusion, although they were not mere tourists either since 'they had entered fully into the life of the Old World. Instead of patronizing tourist hotels they had made excursions to quaint little out-of-the-way restaurants, with the real French atmosphere . . . Not for them the glittering capitals – Paris, Brussels, Rome. They were content with short trips to beautiful historic old towns' (ibid.: 114).

This quotation recalls Fussell's and Buzard's contrast between the adventurous traveller and the vulgar tourist, but it also suggests a third paradoxical alternative, which might be defined as long-term temporary settlement. However, the Fitzgeralds' integration remains questionable, even to the narrator, who regularly relies on self-deprecating humour, as they are victims of their self-assurance and easily taken in by the locals. As noticed by Gerald Kennedy, Fitzgerald is good at 'satirizing [their] own resistance to the foreign' (Kennedy 2002: 120); all his humorous self-narratives are teeming with self-mockery.

Scott's and Zelda's memoirs suggest the comic clash of cultures, the self being the norm, the other the exotic, for example when Scott records the local crooks trying to trick them as far as accommodation is concerned or when both writers recall an unappealing French menu including goat meat (Fitzgerald 1957b: 105–6, 110; Z. Fitzgerald 1992b: 421). In *Tender Is the Night*, Augustine, the French cook, is pictured as a rough and rude woman prone to drinking and reacting violently if upset (Fitzgerald 2018: 298–99). In 'The High Cost of Macaroni', a self-narrative about the Fitzgeralds' 1924–25 winter in Rome, the narrator humorously remembers that on the French Riviera they had to 'support a large and constantly increasing French family who jokingly referred to themselves as [their] "servants"' (Fitzgerald 2022: 343). The descriptions of the locals are derogatory in the Italian self-narrative; they always establish the travellers' sense of superiority. They also provide comic relief, but never display real human interest in otherness. In Italy, after various mishaps, the narrator uses a truncated version of the traveller's proverb and concludes, 'When in Rome – that was the idea! . . . It was when we decided this that the fun began' (ibid.: 352). But this lesson in integration is only a lesson in trickery as they constantly present themselves as the victims of the locals whom they are now ready to trick themselves.

As recalled in 'How to Live on Practically Nothing a Year', the Fitzgeralds first arrived on the Riviera in Hyères. The narrator pictures himself ridiculously trying to speak French: 'Je suis a stranger here, . . . Je veux aller to le best hotel dans le town' (Fitzgerald 1957b: 105). The family and nurse eventually briefly settle at the 'Hôtel du Jardin', the main quality of which being its rate of less than eight dollars a day for the four of them (ibid.: 106). In her memoir entitled '"Show Mr. and Mrs. F. to Number—"', first published in *Esquire* in May–June 1934, Zelda chronologically recalls all the hotels where they stayed from 1920 to 1933. According to her essay, Scott's 'Hôtel du Jardin' was in fact the Grimm's Park Hotel (Z. Fitzgerald 1992b: 421), which points out the fictional, even autofictional, quality of Scott's memoir, which often departs from autobiographical reality and mainly aims at entertaining the reader. In their self-narratives, the Fitzgeralds map the real and the fictional, and this corresponds to what Alexis Tadié notices about travel writing, a literary genre in which 'the frontier between fiction and non-fiction' is often blurred (Tadié 2011: 179).[13]

As described in Scott's self-narrative, once on the Riviera, the travellers first acquire a car (Fitzgerald 1957b: 108), which recalls their Rolling Junk at home, and after being the victims of the locals' deception, they eventually manage to find a villa thanks to the local real estate agent, W.F. King, which involves a comical misunderstanding as they think their search implies mixing with some kind of royalty (ibid.: 111–12). In real life, the Fitzgeralds settled at Villa Marie at Valescure, a residential area of Saint-Raphaël: 'Within an hour we had seen our home, a clean cool villa set in a large garden on a hill above town. It was what we had been looking for all along' (ibid.: 112). In Zelda's novel *Save Me the Waltz*, a *roman à clef* about their life, it becomes 'Les Rossignols' (Z. Fitzgerald 1992a: 76–78). 'Villa Diana', a key location in *Tender Is the Night*, was also partly inspired by the Valescure residence.

A modern couple, the Fitzgeralds have left many photographs, even three silent amateur films,[14] one of them taking place on the French Riviera in 1925–26 (Springer 2012) and showing them wearing the striped sailors' sweaters their friend Gerald Murphy had started to launch as fashionable. These static and motion pictures were visual ways of recounting their travelling experience. In a number of photographs, they are in their villa and even in their car, as during the Rolling Junk road trip, but this time the pictures are not funny but very stylish. These are often family photographs that offer a mixture of relaxation and pose, emphasizing the Fitzgeralds' fashionable glamour, their exciting exotic environment and their social status. Zelda gathered some of them in her scrapbook; there is even a page that focuses entirely on their Saint-Raphaël stay and includes family and landscape pictures.

The F. Scott and Zelda Fitzgerald papers at Princeton University include, among other documents, their busy correspondence and numerous family pictures, but also news articles about them; these constitute a meta-discourse filling in the gaps that their publications, in particular their travel writings, do not always elucidate. In one picture at the Villa Marie, Scott pretends he is playing golf in the driveway (Bruccoli 2003: 118). We know from Zelda's correspondence to Scottie in May 1924 that they had left their daughter at home with her nanny in order to go to Hyères for a few days to play golf while they stayed at the Golf-Hotel there.[15] This was a fashionable activity of the time; before the trip to Europe, Zelda had even been featured in a St Paul newspaper for her golf playing, as shown in her scrapbook, which includes a clipping entitled 'Society Women Compete in Golf Tournament' and a photograph of herself with friend C.O. Kalman in golf outfits outside the clubhouse. Reflecting the new popularity of sports during the 1920s, and golf in particular, in *The Great Gatsby* (1925), Fitzgerald created the secondary character Jordan Baker, a successful golf champion whose honesty on the links is questioned. This new interest in sports meant new clothing styles, especially for women. In *The Cruise of the Rolling Junk*, Fitzgerald contrasts their modern ways with what he calls the 'provincialism' of the South (Fitzgerald 2011: 69). As Zelda's driving outfit is rather badly perceived there, she decides to wear a skirt over it to be accepted in southern hotels (ibid.: 64–71). She, too, recalls the anecdote in her memoir: 'At the O. Henry in Greensboro they thought a man and wife ought not to be dressed alike in white knickerbockers in 1920' (Z. Fitzgerald 1992b: 419). Even at home, otherness is met on the way for these travellers, and their self-narratives emphasize the gap between urban modernity and the deep South of the time. This confrontation between modernity and tradition is also present in Zelda's scrapbook, which proudly shows two pictures of the Fitzgeralds in front of their car during the 1920 road trip: one with the couple in their twin outfits and the other with a self-assured, smiling Zelda, hands in the pockets of her knickerbockers, a relaxed attitude jarring with Southern etiquette in terms of female behaviour.

In these new days of fashionable outdoor sports, swimming was also one of the Fitzgeralds' activities and many of their pictures recall this new entertainment for men, women and children. Much later, during the summer of 1936, when Zelda was away from her daughter and husband due to her mental illness, she would nostalgically recall all the various delightful times when she and Scottie had swum together: 'We have swam [*sic*] together through the translucent blur of the Mediterranean, and in the deep moss greens of Oregon, and over we dived through the waves in a gray slate ocean at Long Beach; we have swam the aquamarine bar of Meadow Brook and I don't know how many more places we've been to to-gether.'[16]

The anaphora 'we have swam', the vivid colours of the water and the distant locations conjure up exotic memories of beauty and happiness. In many pictures, the family can be seen in the water or on the beach, dressed in swimsuits, looking fit and tanned, echoing the Art Deco tourist advertisements of the time celebrating the pleasant climate of the Riviera, and the benefits of sports and outdoor life. During the 1920s, the Riviera beaches became the place to be for a small group of Americans who launched new tourist habits, unlike what had been common with Russian and British visitors of previous decades who had usually left for the northern coast before the summer heat began (Fitzgerald 1957b: 101), thus ending the season on the Riviera. This is also mentioned in *Tender Is the Night* (Fitzgerald 2018: 17, 193).

This novel is dedicated to Gerald and Sara Murphy, whom the Fitzgeralds probably met on the Riviera in 1924, the year Scott published his humorous memoir 'How to Live on Practically Nothing a Year'. The Murphys were wealthy American art lovers, first arrived in Paris in 1921, who were introduced to Cap d'Antibes in 1922 by their friend the musician Cole Porter. They kept coming back; they had their Villa America built, ruled over the Garoupe beach, like Dick Diver in *Tender Is the Night*, and gathered famous artists like Picasso, Hemingway, Dos Passos, Fernand Léger, Stravinsky, Rudolph Valentino and dancers from the Ballets Russes. Thus, in his essay entitled 'Echoes of the Jazz Age', Fitzgerald explains: 'The gay elements of society had divided into two main streams, one flowing towards Palm Beach and Deauville, and the other, much smaller, towards the summer Riviera. One could get away with more on the summer Riviera, and whatever happened seemed to have something to do with art' (Fitzgerald 1965: 15). More humorously, in September 1925, he wrote to John Peale Bishop:

> There was no one at Antibes this summer except me, Zelda, the Valentinos, the Murphys, Mistinguet, Rex Ingram, Dos Passos, Alice Terry, the MacLeishes, Charlie Brackett, Maud Kahn, Esther Murphy, Marguerite Namara, E. Phillips Oppenheim, Mannes the violinist, Floyd Dell, Max and Crystal Eastman, ex-Premier Orlando, Etienne de Beaumont – just a real place to rough it, an escape from all the world. (Turnbull 1964: 359)

Like the Murphys, the Fitzgeralds became fond of the Riviera, thus belying their first impressions of France in 1921. In February 1926, Scott wrote to his editor: 'with our being back in a nice villa on my beloved Riviera (between Cannes and Nice) I'm happier than I've been for years' (Turnbull 1964: 200). Having arrived in Hyères, then settled in Saint-Raphaël, they moved further East along the coast over the years, getting closer to their dear friends, the Murphys. After the 1924–25 winter in Rome and Capri, and spring in Paris, they stayed at Hôtel du Cap at Antibes in August 1925 (Z. Fitzgerald 1992b: 421). Then, following their winter in Paris, they were at the Villa Paquita at Juan-les-Pins in the early spring of 1926, which they eventually left to the Hemingways in early June (Bruccoli 1991: 288) to rent Villa St Louis.[17] Today Villa St Louis has become the Belles Rives hotel at Juan-les-Pins and a plaque in one of its salons celebrates its former literary tenant. In his correspondence, Fitzgerald was enthusiastic about the Riviera and urged his friends to join them. In an undated letter posted from Villa St Louis, he particularly displayed his proud expertise about the region to counter Ludlow Fowler's assumption that they lived 'in a wilderness'. The two-page letter to the latter includes not only written advice but also an annotated map showing a route from Paris to Juan-les-Pins, through Bordeaux and Biarritz, and locating the main sites on the coast: Cannes, Juan-les-Pins, Antibes, Nice, Montecarlo.[18] Here, the analogical image of the coast combines with the written text to induce the foreigner to visit the place the Fitzgeralds knew so well. Zelda's '"Show Mr. and Mrs. F. to Number—"' also suggests they were perfect connoisseurs of the tourist facilities when she nostalgically recalls their occasional stays in glamorous hotels along the coast, in Nice in particular, including the luxury Ruhl in 1924, the Hôtel O'Connor in 1925 and the Beau Rivage in 1929. As usual in her writing, Zelda combines various sensory images that produce an unreal, poetic effect:

> At the Ruhl . . . [d]uring dinner on the terrace, stars fell in our plates. (Z. Fitzgerald 1992b: 421)
>
> In the Hôtel O'Connor old ladies in white lace rocked their pasts to circumspection with the lullabyic motion of the hotel chairs. But they were serving blue twilights at the cafés along the Promenade des Anglais for the price of a Porto, and we danced their tangos and watched girls shiver in the appropriate clothes for the Côte d'Azur. (Ibid.: 423)
>
> Reaching Nice, we went economically to the Beau Rivage, which offered many stained glass windows to the Mediterranean glare. It was spring and was brittly cold along the Promenade des Anglais, though the crowds moved persistently in a summer tempo. We admired the painted windows of the converted palaces on the Place Gambetta. (Ibid.: 426)

The Fitzgeralds were part of the influx of visitors who brought about change along the Riviera with a new season extending to the summer and early autumn.

However, even if they enjoyed France and the Murphys' home was always full of artists of diverse backgrounds, they remained within a predominantly American environment, whether on the Riviera or in Paris, which led them to live in France for several years without mastering the language. Scott could not speak French; Zelda could manage for daily life; and only young Scottie took advantage of these stays and acquired good linguistic skills, which helped her later on to join Vassar College. In 'The High Cost of Macaroni', Fitzgerald is perfectly honest when he declares ironically: 'it seemed a shame to leave France just as we had learned the language – we spoke it now with hardly any accent – that is hardly any French accent' (Fitzgerald 2022: 344). And though he wrote while living in France, he largely ignored his surroundings to focus on the American scene, as in *The Great Gatsby*, mainly written at the Villa Marie in Saint-Raphaël but taking place in New York. And when he did include a French setting, as in *Tender Is the Night* or in various short stories,[19] it remains a backdrop for narratives that focus on American concerns and characters. This echoes Caren Kaplan who contends that distance may offer 'the best perspective on a subject' (Kaplan 1996: 36), and historian Lucy Moore's conclusions about the American expatriates of the 1920s:

> The more insightful among them recognized that, paradoxically, living abroad made it possible to look more clearly at the United States, to better judge and comment on what they had left behind. Their time away actually intensified their Americanness, rather than diluting it, and this became a powerful inspiration for many. Then, too, returning Americans found that they liked being back home – that the familiar had charms more potent than they remembered. (Moore 2008: 240–41)

Geographical decentring seems to have provided the Fitzgeralds with a more acute perception of the home scene for Scott and of her own life as a woman and artist for Zelda.

## Modernists on the Move

In the 1937 essay 'Early Success', Fitzgerald remembers his French Riviera days, which he calls 'the eternal carnival by the Sea', and he concludes: 'life was literally a dream' (Fitzgerald 1965: 62–63). As for Zelda, while visiting Los Angeles in 1927 or long after their Riviera stays when she was a clinic patient due to her nervous collapse, she kept recalling these days in her correspondence as a period of extreme happiness.[20] Moreover, although they could not speak French, after their European stays, many French words and phrases started to pervade their letters, Zelda's in particular. She also evoked these days through her paintings, for example, her 1940s Parisian landscapes,[21] but also a lampshade showing the family, friends and servants on a merry-go-round with familiar places in the background, including Villa Marie and Villa St Louis (Lanahan 1996: 2–3), or *Mediterranean Midi*, fusing together their French and Italian experiences (ibid.: 75). Their time in France, and on the Riviera in particular, remained an enchanted period in their minds, although these times were not devoid of hardships, such as Zelda's affair with French naval aviator Edouard Jozan at Saint-Raphaël in July 1924, her mental breakdown in Paris in April 1930 and Scott's excessive drinking.

Fifty years after the Saint-Raphaël days, helped by Matthew J. Bruccoli and Joan P. Kerr, Scottie offered a visual recollection of her parents' life, including their journeys to Europe, in her publication *The Romantic Egoists: A Pictorial Autobiography from the Scrapbooks and Albums of F. Scott and Zelda Fitzgerald* (1974). Given Zelda's persistent mental illness from 1930 until the end of her life in 1948, this publication can be perceived as what Marianne Hirsch terms a 'postmemory' gesture, describing 'the relationship that later generations or distant contemporary witnesses bear to the personal, collective, and cultural trauma of others – to experiences they "remember" or know only by means of stories, images, and behaviors' (Hirsch 2014: 339). This was Scottie's attempt to grapple with her parents' existence and struggles. Though we know that the Fitzgeralds' time in France was not always truly carefree, the pictures and texts of Scottie's volume suggest a halcyon experience, as do the original scrapbooks, where the photographs and their humorous or enthusiastic comments conjure up a period when the Fitzgeralds embodied youth, creativity and modernism.

As Jean Viviès has argued of Swift's *Gulliver's Travels* (1726), journeys can be apprehended not so much as visited places but through the traveller's capacity to return home (Viviès 2016). For the Fitzgeralds, the final return to America in September 1931 took place within the context of the Great Depression and after Zelda's mental collapse. Financial and health trouble made the return even more difficult and gave rise to an idealization of the Fitzgeralds' French days, notable in their correspondence from then on, especially Zelda's.

In *Travellers, Immigrants, Inmates: Essays in Estrangement*, Frances Bartkowski compares the traveller's mental state and experience to that of a child entering language and social life. She claims that travel places the subject in a situation of dependence and vulnerability resembling that of a young child and, therefore, it involves issues of power and identity: 'this becoming-a-child-again holds the potential for jubilation as much as it does for humiliation at any given moment; this vacillation tells readers much about the relation of writers to their own sense of authority and identity' (Bartkowski 1995: xix). Scott and Zelda reacted differently to expatriation but their writings are in keeping with Bartkowski's theory. Despite their French experience, in his novels, short stories and self-narratives, Scott wrote mainly about Americanness, his sense of identity and his attachment to and disappointment about the American nation. Zelda, meanwhile, focused on the self and her private female experience in both her one published novel and her taxonomic self-narratives. *Save Me the Waltz* focuses on Alabama and David Knight, a young artistic couple closely modelled on the Fitzgeralds, who meet in the Deep South and travel to Paris and the French Riviera, and even reach Naples where Alabama accepts a ballet position.[22] But it is Alabama's aspirations, frustration and despair that are emphasized. In '"Show Mr. and Mrs. F. to Number—"', Zelda details the hotels where they stayed as a couple, and in 'Auction—Model 1934' the objects they collected during their wandering life (Z. Fitzgerald 1992b, 1992c); but what is emphasized is a restless married life, rather than the visited places. The Fitzgeralds' journeys abroad became spaces for Scott's national reflections whereas, for Zelda, they ended in her mental collapse and in writings and paintings betraying a desperate attempt at reaching self-empowerment and self-expression.

Moreover, for both of them, the French experience was not only the meeting of geographic otherness; it was the opportunity, in particular through the Murphys, to be in regular touch with the artistic avant-garde of the time. From then on, neither of them would write (or in Zelda's case paint) the same way. France provided them with an initiatory experience that changed their artistic approach forever and made modernists out of the provincial Americans they had been before despite Scott's initial success with *This Side of Paradise*. Meeting the contemporary writers, painters, musicians and dancers in Paris and on the Riviera who challenged the norms of the times led to their own artistic modernism. In 1925, Fitzgerald wrote to his editor of the developing manuscript of *Tender Is the Night*, 'it is something really NEW in form, idea, structure – the model for the age that Joyce and Stein are searching for, that Conrad didn't find' (Turnbull 1964: 182). From *The Great Gatsby* on, his novels relied on recurring fragmentation, time-shifts and an imagery inspired by modernist painting, just as Zelda acquired a style of her own in her painting, away from her traditional Southern artistic education.

It is worth noting that, although it was mainly written in Saint-Raphaël, *The Great Gatsby* dismisses the European scene. In fact, travelling remains in the background in this narrative mainly focused on American characters leaving their native Midwest for the lure of exciting New York. In fact, it took a further decade for the Fitzgeralds' European travel to find its way into Scott's fiction, as if he needed the return home and some kind of settling to tackle this experience in writing, just as he had drafted *The Cruise of the Rolling Junk* two years after the trip itself. Indeed, it was only once he was back in the United States permanently that he managed to finish *Tender Is the Night*. In the same way, Zelda wrote *Save Me the Waltz* in her American clinic in 1932.

*Tender Is the Night* was published in 1934. It relies on a European setting and was heavily inspired by their travelling experience: it opens on the French Riviera and then takes the reader to Paris, Switzerland and Italy, with a few brief trips back to the United States. Incidentally, a number of short stories with a European setting – mainly Paris and the Riviera[23] – and a travelling motif were published between 1929 and 1931 and worked as preparatory material for the novel. In *Tender Is the Night* as in the stories, it is not so much the European settings that are emphasized as the travelling experience itself: movement reflecting inner uncertainty is at the heart of all of them.

Fitzgerald drew inspiration from his own experience but also from guidebooks to conjure up a European atmosphere. His evocations are not necessarily accurate, as one sees when trying to follow his characters' routes in Paris in 'Babylon Revisited', for example. More than the landscapes and cityscapes, it is the means of transportation and the characters' itineraries that are emphasized. Speed and restlessness are highlighted as most characters come and go on transatlantic liners and then criss-cross Europe ceaselessly. At the beginning of 'The Hotel Child', for example, the narrator introduces the Schwartz family, saying: 'For three years they had lived in hotels – in Paris, Florence, St Raphael, Como, Vichy, La Baule, Lucerne, Baden-Baden and Biarritz' (Fitzgerald 1992: 606). Ships, boats, yachts, trains, cars, taxis, planes, even market wagons, bicycles, funiculars and aquaplanes are emphasized at the expense of the environment itself. Like Gatsby, the characters are 'never quite still' (Fitzgerald 1982: 70). They are compulsive travellers and restless people. Even when they have reached a destination, movement is endless, as in *Tender Is the Night*, when an automobile ballet invades the streets: 'They were twelve, they were sixteen, they were quartets in separate motors bound on a quick Odyssey over Paris' (Fitzgerald 2018: 86). The general impression is always frantic agitation and speed – in contrast with Odysseus' slow and time-consuming Mediterranean wanderings in quest of his return home. Distance and destination do not matter for characters unable to stop and observe their environment.

In the novel, Dick Diver succumbs to his attraction to young starlet Rosemary in a taxi, as if no other place could be in keeping with this disturbing relationship. The city is in motion, blurred by speed and erotic dizziness as Dick gives way to his desire:

> . . . lovers now they fell ravenously on the quick seconds while outside the taxi windows the green and cream twilight faded, and the fire-red, gas-blue, ghost-green signs began to shine smokily through the tranquil rain. It was nearly six, the streets were in movement, the bistros gleamed, the Place de la Concorde moved by in pink majesty as the cab turned north. (Fitzgerald 2018: 83)

The Parisian cityscape here is all colours and movements, as in an impressionist painting, far from a realistic representation of what the Fitzgeralds experienced of Paris as visitors.

Travelling turns out to be inadequate to heal and soothe those troubled characters' minds. Obsessed by gossip about Rosemary's behaviour with a young man on a train, Dick keeps imagining an intimate travelling scene that contradicts her naïve innocence:

> – Do you mind if I pull down the curtain?
> – Please do. It's too light in here. (Fitzgerald 2018: 100)

As in an obsessive process, those words recur repeatedly in the narrative (Fitzgerald 2018: 100, 101, 113, 191), thus conjuring up both Rosemary's licentious behaviour and Dick's uneasy mind. The train scene even reaches a climax of anguish for Dick when he thinks of it while in 'a wagon-lit' with his wife (ibid.: 190): 'It was worse with his eyes shut for it gave a rhythm of finding and losing, finding and losing; but so as not to appear restless he lay like that until noon' (ibid.). The *mise en abyme* of the train scene evokes a disgust and sickness that encapsulate the character's torment and loss, thus heralding his doom. As for the wife, Nicole, Dick's former psychiatric patient, travelling provides no solace either. In her unique stream of consciousness in the novel (ibid.: 181–85), she remembers frantic travelling from Australia to Algeria, but the images are rather unpleasant, conjuring up discomfort, prostitution and poverty:

> We travelled a lot that year – from Woolloomooloo Bay to Biskra. On the edge of the Sahara we ran into a plague of locusts and the chauffeur explained kindly that they were bumble-bees. The sky was low at night, full of the presence of a strange and watchful God. Oh, the poor little naked Ouled Naïl;[24] the night was noisy with drums from Senegal and flutes and whining camels, and the natives pattering about in shoes made of old automobile tires. (Fitzgerald 2018: 183)

Working on his novel, Fitzgerald drew a chart[25] comparing Zelda's and Nicole's psychiatric conditions, thus feeding fiction from real life. Like Dick and Nicole Diver, the Fitzgeralds had been to Algeria themselves in February 1930 in an effort to calm Zelda, who was already on the edge of a mental collapse, which eventually occurred in June and led to her entering various clinics in France, then in Switzerland, until their return to the United States in September 1931. The trip was no real success, despite a few funny pictures in the desert with camels, as Zelda was too stressed because she was missing her ballet lessons in Paris. The evocation of Algeria and Nicole's nervous instability in *Tender Is the Night* echoes Zelda's reference in '"Show Mr. and Mrs. F. to Number—"': 'The world crumbled to pieces in Biskra' (Z. Fitzgerald 1992b: 428). Here, Zelda's words are no comment on the North African town's decay; they actually discretely refer to her own nervous condition. In the same way, their next travelling step implies her painful state and clinic internment in a veiled manner: 'Then Switzerland and another life' (ibid.: 428).

For the Fitzgeralds, Switzerland came to embody Zelda's health trouble, as suggested in her memory of the Lausanne Palace and its 'chic fugitives from life and death' (Z. Fitzgerald 1992b: 428). In Scott's fiction, Switzerland always represents the final stage of personal collapse, as underlined in similar passages in *Tender Is the Night* (Fitzgerald 2018: 279) and 'The Hotel Child': 'this corner of Europe does not draw people; rather, it accepts them without too many inconvenient questions – live and let live. Routes cross here – people bound for private *cliniques* or tuberculosis resorts in the mountains, people who are no longer *persona grata* in Italy or France' (Fitzgerald 1992: 598).

As is clear in 'One Trip Abroad', Switzerland stands for the end of all journeys, the final step of an entropy process: 'Having visited North Africa, Italy, the Riviera, Paris and points in between, it was not surprising that eventually the Kellys should go to Switzerland. Switzerland is a country where very few things begin, but many things end' (Fitzgerald 1992: 594). Like the Kellys in that story, the Fitzgeralds experienced the end of their carefree travelling years in Europe with the Swiss episode, but their journeys abroad kept inspiring their fiction and essay writings.

In *The Sun Also Rises*, from 1926, Ernest Hemingway has his narrator Jake Barnes exclaim: 'going to another country doesn't make any difference. I've tried all that. You can't get away from yourself by moving from one place to another. There's nothing to that' (Hemingway 1970: 11). For the Fitzgeralds, France never made them 'get away from [themselves]', especially as they remained mainly among American friends – including Hemingway himself. But it changed their artistic approach as it opened their horizons and knowledge, and enabled them, through their works, to enter the modernist age fully. The travelling experience transformed them as artists, even though they remained rather impervious to foreign customs and languages, and never really mingled with the locals. Foreignness and travelling challenged their artistic skills rather than their daily life. Thus, when they returned home in 1931, they were completely changed as artists; their texts and her paintings had become emblematic of modernity.

## Conclusion

The Fitzgeralds were always on the move. They were travellers, if not exactly travel writers, and their pictures, letters, life-narratives and novels are teeming with evocations of their life abroad, especially in France. In the end, their works are a kind of modernist transmedia storytelling. They remained Americans at heart, but their French experience transformed them as artists and remained in their minds forever, long after departure and after Zelda's mental collapse. Among the many documents bequeathed by their daughter Scottie to the Special Collections of Firestone Library at Princeton University, one finds multiple traces of their stays in France: pictures, passports, French identity papers for foreigners, insurance contracts, an international driving license, a touring club membership card, French doctors' examinations and prescriptions, casino and theatre tickets, a typed bouillabaisse recipe[26] recalling Nicole Diver's interest in cooking in *Tender Is the Night* (Fitzgerald 2013: 20) and more. It is perhaps Zelda's address book that offers the best testimony of their French experience.[27] Echoing Zelda's taxonomic essays, it is a succession, in French and English, of various contacts, services, places and goods to be found in the US, but mainly in Paris and on the Riviera: addresses for servants (*'femme de ménage'*, *'lingère'*, *'cuisinière'*, *'femme de chambre'*), cleaners, dressmakers, nurses, lawyers, dentists, doctors, ballet teachers, typists, actors, restaurants, a 'laiterie', a gym, hats, shoes, toys, 'perfume from Cannes', and friends and acquaintances like the Murphys, Ernest Hemingway, Sylvia Beach, James Joyce, John Dos Passos, Carl Van Vechten, King Vidor and Edith Wharton. This address book, in its own way, is a traveller's narrative as it tells of a life among French people and English-speaking artists during the advent of modernism, the life of the true modernists France enabled the Fitzgeralds to become.

**Elisabeth Bouzonviller** is Professor of American literature at Jean Monnet University in St Etienne, France. She is the author of *Francis Scott Fitzgerald, écrivain du déséquilibre* (Belin, 2000) and *Louise Erdrich: Métissage et écriture, histoires d'Amérique* (Presses Universitaires de St Etienne, 2014). She regularly contributes to the *F. Scott Fitzgerald Review* and *Newsletter*. She co-edited *Mémoires, traces, empreintes* (Orbis Tertius, 2020) and *F. Scott Fitzgerald: Tender Is the Night* (Ellipses, 2022), and edited *'Home Sweet Home': Places of Belonging in Anglophone Narratives* (Editora, 2022). She also published a preface for a new translation of *The Great Gatsby* (Gallmeister, 2023).

## Notes

1. Author's translation from the original French: 'tout récit est un récit de voyage – une pratique de l'espace'.
2. Scrapbooking, which had begun in the United Kingdom in the nineteenth century, was also fashionable in the United States during the nineteenth and twentieth centuries. For example, Mark Twain travelled with his scrapbooks, which he fed on the way with souvenirs and photographs. Both Scott and Zelda Fitzgerald had their own scrapbooks in which they collected pictures with captions, newspaper clippings, tickets, menus, invitations, telegrams, postcards, ribbons, notes, drawings, dried flowers and memorabilia of all sorts to mark events at home and abroad. Their daughter, Scottie Fitzgerald Smith, bequeathed them all to Princeton University Special Collections in 1950. For Scott, this practice was certainly part, like his *Ledger*, of his desire to record events not only as personal memories but also with a view to leaving elements about himself contributing to a future author's biography, especially as he claimed that 'There never was a good biography of a good novelist. There couldn't be. He is too many people if he's any good' (Bruccoli 1978: 159). Scottie indicates that the 'scrapbooks stop at 1936,

when [her] father moved to North Carolina from Baltimore and put them temporarily in storage' (Bruccoli 2003: ix). These were difficult times for both of them, Zelda being regularly a psychiatric clinic patient and Scott afflicted with alcoholism. financial problems and a lack of creative inspiration.

3. Some are to be found at Princeton University Special Collections, including passports, French identity papers, insurance policies, a touring club membership card or an international driving licence (F. Scott Fitzgerald Papers, C0187, Box 51, Folders 1, 5, 10 and 11).
4. Deborah Pike describes Zelda as follows: 'Well known as a flirt and social butterfly' (Pike 2017: 52).
5. The Fitzgeralds' first sojourn in Europe took place from May to July 1921. The second stay lasted from May 1924 to December 1926. Their next two visits lasted from April to early October 1928 and from March 1929 to September 1931. Scott left France on 30 January 1931 to attend his father's funeral in the United States while Zelda stayed at Prangins clinic in Switzerland and he came back to Europe in early March. Eventually, they returned permanently together to the US in September 1931.
6. The James quotation is as follows: 'Who shall say, at the rate things are going, what is to be "near" home in the future and what is to be far from it? . . . The globe is fast shrinking, for the imagination, to the size of an orange that can be played with; the hurry to and fro over its surface is that of ants when you turn up a stone . . .' (James 1898: 650).
7. In *The Education of Henry Adams*, considering the 1200–1900 period, the narrator develops an acceleration theory applied to society (Adams 1973: 489–98). He claims that '[t]he movement from unity into multiplicity . . . was unbroken in sequence, and rapid in acceleration' (ibid.: 498).
8. However, Fussell was excessively pessimistic, as travel literature has actually had wonderful years since the Second World War. Thus, in *The Travel Writing Tribe*, Tim Hannigan embarks on a quest to answer the nagging question 'Is travel writing dead?' (Hannigan 2021: 6, 13), and the answer is obviously negative, as the authors he enjoyed in his youth – Eric Newby, Paul Theroux, Nicholas Crane or even the controversial Bruce Chatwin (ibid.: 3–4, 314–19) – the ones whose archives he studies after years of admiration – Wilfred Thesiger (ibid.: 83–97) and Patrick Leigh Fermor (ibid.: 267–83) – and the ones he meets are all successful post-Second World War authors. In his work, he interviews travel writers from various generations and there is no doubt that these – Philip Marsden, Nicholas Jubber, Colin Thubron, Nick Danziger, Robin Hanbury-Tenison, Dervla Murphy, Rory MacLean, Sara Wheeler, Samanth Subramanian, William Dalrymple and Kapka Kassabova – have escaped the plagues of tourism criticized by Fussell. Finally, countering Hannigan and Fussell's fear that travel writing might have become an obsolete genre, Wheeler offers an optimistic assertion: 'I believe that there will be lots more men and women coming after me who will write brilliant things about observing countries' (ibid.: 203).
9. Augé notices that the 'supermodern world' has experienced a 'change of scales' to be noticed in 'the accelerated circulation of passengers and goods' (Augé 1995: 33–34), eventually bringing about 'the contraction of the planet' (ibid.: 119).
10. Princeton University Special Collections: F. Scott Fitzgerald Additional Papers, C0188, Box 4, Folder 9.

11. Author's translation from the original French: 'cette capacité à prendre distance d'avec soi, ses repères, ses convictions, sa vision du monde, pour aller à la rencontre d'autrui'.
12. Author's translation from the original French: 'le décentrement marque l'intérêt pour l'altérité et la tentative de ne pas examiner autrui avec un regard ego- ni ethno-centré'.
13. This uncertain frontier is precisely one of the topics at stake when Hannigan interviews travel writers in *The Travel Writing Tribe*.
14. One film shows the Fitzgeralds with their baby daughter, another their sitting and chatting on a Mediterranean terrace with friends, and the third captures Scott at a writing table in a garden, which may have been located at Villa Marie in Saint-Raphaël.
15. Princeton University Special Collections: *Zelda Fitzgerald Papers*, C0183, Box 4, Folder 2.
16. Princeton University Special Collections: *Zelda Fitzgerald Papers*, C0183, Box 4, Folder 27.
17. This and subsequent quotations are from Princeton University Special Collections: *F. Scott Fitzgerald Additional Papers*, C0188, Box 4, Folder 22.
18. In his letter, Fitzgerald advises Fowler to drive from Biarritz 'through Pau, Carcassonne, Nimes, ect [*sic*]' since 'by auto its [*sic*] wonderful' whereas 'from Biaritz [*sic*] to Marseille by train is two terrible days'.
19. See 'The Swimmers', 'One Trip Abroad', 'The Hotel Child', 'Babylon Revisited', 'Not in the Guide Book' and 'Image on the Heart', for example.
20. For example, in her letters to Scottie, she wrote: 'I'd like more than anything to be touring from Paris to the Riviera along the white roads' (January–February 1927 from Los Angeles, *Zelda Fitzgerald Papers*, C0183, Box 4, Folder 4); 'I have been in the pool here and it was lovely but not so nice as swimming in the ocean at Antibes' (March 1927 from Los Angeles, *Zelda Fitzgerald Papers*, C0183, Box 4, Folder 11); 'Have a good time and build the palace of the Popes at Avignon in the sand for me. Do you remember the faces we made from wet sand at Antibes?' (August 1931 from Prangins clinic, *Zelda Fitzgerald Papers*, C0183, Box 4, Folder 21); 'the Mediterranean always was a heavenly spot to us' (*c*. February 1943, *Zelda Fitzgerald Papers*, C0183, Box 4, Folder 43).
21. *Notre Dame Cathedral*, *Place de l'Opera*, *The Pantheon and Luxembourg Gardens*, *Arc de Triomphe* and *Madeleine Aperitifs* (Lanahan 1996: 57–61).
22. Zelda herself never responded to an offer from Madame Sedova to join the San Carlo Opera Ballet Company in Naples in 1929 (Wagner-Martin 2004: 125).
23. 'Not in the Guide Book' (1925), 'The Swimmers' (1929), 'The Rough Crossing' (1929), 'One Trip Abroad' (1930), 'The Hotel Child' (1931) and 'Babylon Revisited' (1931).
24. Ouled Naïls were young Berber women trained as belly dancers who often turned out to be prostitutes. Zelda also mentions them in '"Show Mr. and Mrs. F. to Number—"': 'The Ouled Naïls were very brown and clean-cut girls, impersonal as they turned themselves into fitting instruments for sex by the ritual of their dance' (Z. Fitzgerald 1992b: 428).
25. Princeton University Special Collections: *F. Scott Fitzgerald Papers*, C0187, Box 7, Folder 1.
26. Princeton University Special Collections: *Zelda Fitzgerald Papers*, C0183, Box 6, Folder 2.
27. Princeton University Special Collections: *Zelda Fitzgerald Papers*, C0183, Box 6, Folder 1.

## References

Adams, Henry. 1973 (1918). *The Education of Henry Adams*. Boston: Houghton Mifflin Harcourt.

Augé, Marc. 1995 (1992). *Non-Places: Introduction to an Anthropology of Supermodernity*, trans. John Howe. London: Verso.

Bartkowski, Frances. 1995. *Travellers, Immigrants, Inmates: Essays in Estrangement*. Minneapolis: University of Minnesota Press.

Bruccoli, Matthew J. (ed.). 1978. *The Notebooks of F. Scott Fitzgerald*. New York: Harcourt Brace Jovanovich/Bruccoli Clark.

———. 1991. *Some Sort of Epic Grandeur: The Life of F. Scott Fitzgerald*. London: Cardinal.

Bruccoli, Matthew J., Scottie Fitzgerald Smith, and Joan P. Kerr (eds). 2003 (1974). *The Romantic Egoists: A Pictorial Autobiography from the Scrapbooks and Albums of F. Scott and Zelda Fitzgerald*. Columbia: University of South Carolina Press.

Buzard, James. 1993. *The Beaten Track: European Tourism, Literature, and the Ways to Culture, 1800–1918*. Oxford: Clarendon Press.

Casanova, Pascale. 2014. 'Paris, méridien de Greenwich de la littérature', in Christophe Charle and Daniel Roche (eds), *Capitales culturelles, capitales symboliques: Paris et les expériences européennes (XVIII^e–XX^e siècles)*. Paris: Éditions de la Sorbonne, pp. 289–96.

Cossu-Beaumont, Laurence. 2023. *Deux agents littéraires dans le siècle américain: William et Jenny Bradley, passeurs culturels transatlantiques*. Paris: ENS Éditions.

De Certeau, Michel. 1990. *L'Invention du quotidien*. Paris: Gallimard.

Fitzgerald, F. Scott. 1948 (1920). *This Side of Paradise*. New York: Scribner's.

———. 1957a (1924). 'How to Live on $36 000 a Year', in Arthur Mizener (ed.), *Afternoon of an Author: A Selection of Uncollected Stories and Essays*. New York: Scribner's, pp. 87–99.

———. 1957b (1924). 'How to Live on Practically Nothing a Year', in Arthur Mizener (ed.), *Afternoon of an Author: A Selection of Uncollected Stories and Essays*. New York: Scribner's, pp. 100–16.

———. 1965. *The Crack-Up with Other Pieces and Stories*. London: Penguin.

———. 1979. *The Price Was High: The Last Uncollected Stories of F. Scott Fitzgerald*. London: Quartet.

———. 1982 (1925). *The Great Gatsby*. Harmondsworth: Penguin.

———. 1992. *The Short Stories of F. Scott Fitzgerald*. London: Abacus.

———. 2011 (1924). *The Cruise of the Rolling Junk*. London: Hesperus Press.

———. 2018 (1934). *Tender Is the Night*. London: Penguin English Library.

———. 2022 (1954). 'The High Cost of Macaroni', in Jim West (ed.), *The Last Kiss*. Cambridge: Cambridge University Press, pp. 343–55.

Fitzgerald, Zelda. 1992a (1932). *Save Me the Waltz*, in Matthew J. Bruccoli (ed.), *Zelda Fitzgerald: The Collected Writings*. London: Little, Brown, pp. 9–196.

———. 1992b (1934). '"Show Mr. and Mrs. F. to Number—"', in Matthew J. Bruccoli (ed.), *Zelda Fitzgerald: The Collected Writings*. London: Little, Brown, pp. 419–31.

———. 1992c (1934). 'Auction—Model 1934', in Matthew J. Bruccoli (ed.), *Zelda Fitzgerald: The Collected Writings*. London: Little, Brown, pp. 433–41.

Fussell, Paul. 1980. *Abroad: British Literary Travelling between the Wars*. Oxford: Oxford University Press.

Hannigan, Tim. 2021. *The Travel Writing Tribe: Journeys in Search of a Genre*. London: Hurst.
Hemingway, Ernest. 1970 (1926). *The Sun Also Rises*. New York: Scribner's.
Hirsch, Marianne. 2014. 'Presidential Address 2014: Connective Histories in Vulnerable Times', *PMLA* 129(3): 330–48.
Hynes, Samuel. 1982. 'Review of *Abroad*', in *Journal of Modern History* 54(2): 366–69.
James, Henry. 1898. 'The Story-Teller at Large: Mr. Henry Harland', in *Fortnightly Review* 63(374): 650–54.
Joyce, James. 1922. *Ulysses*. Paris: Shakespeare and Company.
Kaplan, Caren. 1996. *Questions of Travel: Postmodern Discourses of Displacement*. Durham, NC: Duke University Press.
Kennedy, Gerald. 2002. 'Fitzgerald's Expatriate Years and the European Stories', in Ruth Prigozy (ed.), *The Cambridge Companion to F. Scott Fitzgerald*. Cambridge: Cambridge University Press, pp. 118–42.
Korte, Barbara. 2000. *English Travel Writing from Pilgrimages to Postcolonial Explorations*. New York: St Martin's Press.
Lanahan, Eleanor (ed.). 1996. *Zelda: An Illustrated Life; The Private World of Zelda Fitzgerald*. New York: H. N. Abrams.
Mellow, James R. 1985. *Invented Lives: F. Scott and Zelda Fitzgerald*. London: Souvenir Press.
Moore, Lucy. 2008. *Anything Goes: A Biography of the Roaring Twenties*. London: Atlantic.
Peraldo, Emmanuelle (ed.). 2016. *Literature and Geography: The Writing of Space throughout History*. Newcastle-upon-Tyne: Cambridge Scholars Press.
Pike, Deborah. 2017. *The Subversive Art of Zelda Fitzgerald*. Columbia: University of Missouri Press.
Regard, Frédéric. 2009. *British Narratives of Exploration: Case Studies on the Self and Other*. London: Routledge.
Springer, Mike. 2012. 'Rare Footage of Scott and Zelda Fitzgerald from the 1920s', *Open Culture*, 7 November. Retrieved 2 July 2024 from https://www.openculture.com/2012/11/rare_footage_of_scott_and_zelda_fitzgerald_from_the_1920s.html.
Stein, Gertrude. 2003 (1940). *Paris France*. London: Peter Owen.
Tadié, Alexis. 2011. 'The Practice of Travel-Writing: From Travel to Theory', in Madeleine Descargues-Grant (ed.), *Récit de voyage et encyclopédie*. Valenciennes: Presses Universitaires de Valenciennes, pp. 171–86.
Turnbull, Andrew (ed.). 1964. *The Letters of F. Scott Fitzgerald*. London: Bodley Head.
Vandamme, Pierre-Étienne. 2017. 'Qu'est-ce que le décentrement? Moralité individuelle et justice sociale', *Ethica* 21(1): 167–202.
Viviès, Jean. 2016. *Revenir/devenir: Gulliver ou l'autre voyage*. Paris: Éditions rue d'Ulm.
Wagner-Martin, Linda. 2004. *Zelda Sayre Fitzgerald: An American Woman's Life*. New York: Palgrave Macmillan.

## Archives

*F. Scott Fitzgerald Papers, 1897–1944*. Special Collections, Firestone Library, Princeton University, C0187, Box 7, Folder 1; Box 51, Folders 1, 5, 10 and 11.

*F. Scott Fitzgerald Additional Papers, 1909–2013*. Special Collections, Firestone Library, Princeton University, C0188, Box 4, Folders 9 and 22.

*F. Scott Fitzgerald's Scrapbook, 1896–1935*. F. Scott Fitzgerald Papers, Special Collections, Firestone Library, Princeton University. Retrieved 24 February 2025 from https://findingaids.princeton.edu/catalog/C0187_c03402?onlineToggle=false.

Matthew J. & Arlyn Bruccoli Collection of F. Scott Fitzgerald. *Ledger, 1919–1938*. University of South Carolina. Irvin Department of Rare Books and Special Collections, Columbia, SC, 29208. Retrieved 20 June 2025 from https://digital.tcl.sc.edu/digital/collection/fitz/id/4.

*Zelda Fitzgerald Papers, 1917–1997*. Special Collections, Firestone Library, Princeton University, C0183, Box 4, Folders 2, 4, 11, 21, 22, 27 and 43; Box 6, Folders 1 and 2.

*Zelda Fitzgerald's Scrapbook, circa 1905–1926*. Zelda Fitzgerald Papers. C0183, Special Collections, Firestone Library, Princeton University. Retrieved 24 February 2025 from https://findingaids.princeton.edu/catalog/C0183_c0280?onlineToggle=false.

# CHAPTER 4
# Terror, Pity, Love
## *The Trials and Tribulations of Annemarie Schwarzenbach and Ella Maillart on the Road to Kabul*

Julia Szołtysek

### Introduction: The Journey and Its Titles

When gathering material for *The Heroine's Journey*, which later became her seminal work, Maureen Murdock asked Joseph Campbell how the woman's journey related to the quest of the hero. The answer that he gave stunned Murdock, who recalls that she found it 'deeply unsatisfying' (Murdock 2020: 2). Campbell said:

> In the whole mythological tradition the woman *is there*. All she has to do is realize that she's the place that people are trying to get to. When a woman realizes what her wonderful character is, she's not going to get messed up with the notion of being pseudo-male. (Ibid.; original emphasis)

What shocked Murdock was not so much the narrow-mindedness of Campbell's response as his obliviousness to the mere possibility of there being any other options available to women. On top of that, Murdock also found alarming the ease with which Campbell consigned women to the role of sessile Penelopes, waiting – and weaving, and weeping – for the return of their heroes. Baffled and disappointed, Murdock set about mapping the female quest, placing the woman right at the forefront of the journey – as the traveller, not travellee, the subject, never object, and the fighter, not the one fought over or fought for.

That was in 1981, only six years before Campbell's passing and over thirty after the first publication of *The Hero with a Thousand Faces*; Campbell was an esteemed and influential scholar and Murdoch an ambitious novice and his former student. Curiously enough, Murdock's 1981 exchange with Joseph Campbell bears some resemblance to a conversation between Ella Maillart, the Swiss-Danish traveller, reporter and writer, and Carl Gustav Jung, to whom Maillart had paid quite a casual visit before she set off on a trip to Afghanistan with Annemarie Schwarzenbach in 1939:

> I called on C.G. Jung, hoping (very foolishly) that he could give me a key to the mentality of the so-called primitives. I offered him one of my books. He looked at it and asked: 'Why do you travel?' 'To meet those who know how to live peacefully' was the first answer that came to my lips. But the great man looked at me with suspicion; did I look like a restless lunatic who wants to be cured? (Maillart 2013: 10)

It is somewhat unsettling to read these two passages, which, though separated by some forty years of turbulent history, seem to strike quite similar chords. The question of women's travels and generally of women-as-travellers continues to be a cause for alarm or at least for the furrowing of the brows (spare a thought for Freud's notorious frown); luckily enough, the stories of women like Ella Maillart, Annemarie Schwarzenbach and, in fact, staggering numbers of other women travellers annul much of the stinging capital of such attitudes. None of them stopped to take heed or wavered in her resolution to travel, be it outwards or inwards, brought down by instances of advice-giving or gaslighting. In fact, Schwarzenbach and Maillart loom as discordant figures, defying limitations and conventions, appropriating instead the terrains traditionally designated as male-only. What is more, their movements and transitions transcend superficial divisions and categories – and it is predominantly these traversals that I wish to focus on in this chapter. While Maillart seems to be embracing life and adventure and in this sense embarks on the journey with jouissance and enthusiasm, for Schwarzenbach the trip appears to promise numerous chances to disembark completely in pursuit of self-annihilation, inviting risk and offering her opportunities to gradually give up on her will to live. While engagement with danger has marked (and marred) Schwarzenbach's entire existence, for Maillart the recklessness displayed by her travelling companion has all too often proven a cause for exasperation and irritation, especially given that in Maillart's moral and ethical code, life and self-respect occupied a high and deeply cherished place.

In an attempt to follow up on these threads, I wish to shed light on modes of departure as deliverance that Annemarie Schwarzenbach and Ella Maillart negotiated and that marked them off as, at once, representatives of the 'leisure classes' of old and harbingers of the tragic heroes of today's grey zones. Both women kept journals during the trip and published their accounts; for Schwarzenbach it was actually a way of acquiring extra funds for the journey –

she regularly sent out her texts in the form of feuilletons and reports to various magazines and journals, which were collected in one volume and published as a complete travelogue under the title *All the Roads Are Open* only as recently as 2000 (in German; the first English translation appeared in 2011). Maillart's *The Cruel Way: Switzerland to Afghanistan in a Ford, 1939* appeared in 1947, and its most recent edition was published in 2013 with a foreword by Jessa Crispin.

It is worth looking more closely at the titles of these two works. While both are direct quotations from the texts, they refer specifically to only one of the travellers – Schwarzenbach, yielding significant insights into her general disposition throughout the trip. When crossing the Khyber Pass, Schwarzenbach was asked by British customs officers for her passport and car documents. She relates how befuddled they were upon finding out that she had never been there before, given how extensive her travels had been. Schwarzenbach does not feel fit to explain herself to them; rather, she seems to withdraw from the entire scene and give in to her wistfulness, noting mournfully that 'certainly, all the roads are open, and lead nowhere, nowhere' (Schwarzenbach 2018: 108). This is just one of the many bursts of despair and regret that she committed along the way – very symptomatic, though. Contrary to what the phrase on its own might appear to mean, this was not an enthusiastic exclamation about the journey and the sense of joy an empty road at one of the world's great crossroads might give to a driver; it was a pronouncement of resignation and sorrow that plagued Schwarzenbach not only during this particular trip, but throughout her whole life, tormenting her and frequently sending her over the edge. In this sense, it does actually correlate with Maillart's choice of title for her work. Maillart, a pragmatic optimist, was experiencing quite profound emotions caused by the circumstances and mused about explaining them to Schwarzenbach in an attempt to prove to her that one can 'bring forth from my being pure and unconditioned joy' (Maillart 2013: 72). By this stage, however, she knew that these thoughts would be lost on her companion, and moved by Schwarzenbach's struggle, she wondered 'why again and again she chose the complicated, cruel way of hell. Could it be that she preferred it to an easier mode of living? Did she believe it was a quick way of exceeding the limitations of her individuality?' (ibid.). Thus, 'the cruel way' becomes a definition of how Schwarzenbach proceeded in the world, how she went about her life; at the same time it rings very true with regards to the trip itself and to its larger, even mythological, implications. Bodily movement, while not capable of saving Schwarzenbach from herself entirely, still provided a shield against emotional stasis in that it occupied the senses and busied the muscles as well as the mind, not allowing the traveller to slide into despair and immobilizing self-pity while on the move. In fact, the 'crueller' the way, the better it proved for keeping Schwarzenbach in check, which is exactly what Maillart's work is about. Curiously enough, it tells us next to nothing about Ella Maillart, the author and the traveller, placing the focus quite definitively on Schwarzenbach, who emerges as the main protagonist and the sole subject of both of these accounts. The practice, though rather unpremeditated on the part of Maillart, might distantly echo the devices applied by James Boswell in *The Journal of a Tour to the Hebrides with Samuel Johnson, LL.D.* (1785). This work's unquestioned hero is Johnson, and the entire text might be read as a 'prolegomenon' of sorts to Boswell's later biography of Johnson. Undoubtedly, the decision to write a travel journal or memoir where the focus is on a protagonist other than the writer is unusual and curious, one that seems to go against the old adage of making a travel account a boastful record of one's own adventures and triumphs or mishaps and sufferings.

## A Belated Icon: Schwarzenbach's Twenty-First-Century Status

In recent years, these two women travel writers have been enjoying a renewed interest, in particular Schwarzenbach, and the rediscovery has been fuelled also by fashion and popular culture. Schwarzenbach surely makes for a fitting hero(ine) – her gender ambiguity and strikingly androgynous appearance, along with a penchant for cross-dressing, single her out as a kind of belated icon. In fact, her photographic appeal had already been eagerly explored by many of her peers and companions, Maillart included. In 2018 the fashion house Givenchy presented a collection inspired by Schwarzenbach, with models sporting silk shirts and high-waisted trousers and shorts, complemented by hairstyles reminiscent of Schwarzenbach. The 2018 fashion show turned out to be a celebration of inclusivity and difference and a peculiar tribute to Schwarzenbach, though most of the 'ugly' parts of her story had been painted over, perhaps understandably. Of these there were plenty – Annemarie Schwarzenbach did not in the least bit feel like an 'icon' or a role model of any sort; throughout her whole life she battled depression, drug addiction, anxiety and neurosis that led her to several suicide attempts and multiple stays in recovery clinics. Her hypersensitivity meant that she could never turn a blind eye to any instance of oppression, which in turn thrust her into a deep conflict with her family and did not earn her many supporters among the social circles within which the Schwarzenbachs moved – to put it bluntly, Alfred and Renne Schwarzenbach, Annemarie's parents, were outspoken Nazi sympathizers who staunchly opposed their daughter's friendship with the family of the avowedly anti-Nazi author Thomas Mann, in particular his children, the artistically minded Klaus and Erika, and pressurized her to cut ties with other acquaintances they deemed improper. In a way, though, these circumstances make her a figure that many young people today can relate to and identify with, which perhaps explains the recent popular attention she has received. On top of that, Schwarzenbach was a highly educated and an extremely well-read person (with a doctorate in history from Zurich and Paris) whose talents included first and foremost writing, both creative and journalistic, photography, and driving, the latter being one of the passions that brought her close to Maillart, also an avid driver.

A more widespread scholarly attention came with English translations – in 2011 Isabel Fargo Cole translated *All the Roads Are Open* (Seagull Books), and in the same year Lucy Renner Jones's translation of *Lyric Novella* appeared, also with Seagull Books; shortly thereafter, in 2012, Renner Jones translated *Death in Persia*, which was published in 2013 again by Seagull Books. All the while numerous blogs, websites and many Pinterest and Instagram profiles emerged devoted to Schwarzenbach, some of which are still up and running, for example *Strange Flowers*, run by James J. Conway, or *The Ravaged Angel*. Importantly, there have also been some substantial academic publications on Annemarie Schwarzenbach, most notable of which are Sofie Decock's and Uta Schaffers's 2011–12 series of articles (in German) about Schwarzenbach's Asian and African travel journals. One of Decock's articles appeared in English in the 2011 issue of *Women in German Yearbook* published by the University of Nebraska Press. The 2016 *Routledge Companion to Travel Writing*, edited by Carl Thompson, includes a shorter and more survey-like entry about Schwarzenbach by David Farley, who interrogates Schwarzenbach's works translated into English from the vantage of modernist travel writing, along authors such as Aimé Césaire and Blaise Cendrars. More recently, Nicole Maruo-Schröder, Sarah Schäfer-Althaus and Uta Schaffers's edited collection *Traveling Bodies: Interdisciplinary Perspectives on Traveling as an Embodied Practice* (2023) includes a chapter by Sofie Decock focused on Schwarzenbach. To date, there is no complete English-language biography of Schwarzenbach. The lack is quite painful because not only are there no original biographical works in English, translations of the ample German, Swiss or French ones are also hard to find. A curious case is that of the immensely popular Italian author Melania Mazzucco, whose immersive fictionalized biography of Schwarzenbach, *Lei cosi amata* (2000), has been translated into at least ten languages, including Polish (*Tak ukochana*; translation by Monika Woźniak, 2006), but never into English. In other media, in 2001 Schwarzenbach's Afghan trip with Maillart was turned into an art-house film, *The Journey to Kafiristan*, directed by Fosco and Donatello Dubini, with Jeanette Hain as Annemarie Schwarzenbach and Nina Petri as Ella Maillart, which however received only a lukewarm critical and audience reception. Yet another popular culture reference to Schwarzenbach, one perhaps not quite so obvious, is Suzanne Vega's 2011 one-woman show *Carson McCullers Talks about Love* in which Vega uses the writer's own words to sing about McCullers's life and experiences, with the author's feelings for Schwarzenbach making up one track in the collection – 'Song of Annemarie'.

## Schwarzenbach and the Heroine's Journey

There surely still remains a lot to explore with regards to Schwarzenbach's life and *oeuvre*, but the primary sources are probably already exhausted – it is unlikely that there will be any new publications by Schwarzenbach because sadly all her letters, journals and other unfinished projects, which may have been prime examples of writing on the move, were destroyed by her mother Renee shortly after Schwarzenbach's untimely death. Renee felt they shed unfavourable light on the family – which they very probably did, because Schwarzenbach was a staunch opponent and merciless critic of Hitler and Nazism, whereas her family, and her mother in particular, openly supported the Führer. This might also explain why Schwarzenbach felt the urge to flee her family home and its atmosphere – she could not accept her parents' political sympathies, especially given that the circle of her closest friends included many people of Jewish heritage and outspoken critics of the regime, including Marianne Breslauer and the Manns.

However, the strange and somewhat incongruent aspect is the fact that her mother was actually openly bisexual, involved in an affair with the opera singer Emily Krueger and others; what is more, Renee raised Annemarie as a boy, styling her hair boyishly and dressing her in boys' clothes. Her daughter's queerness was no secret to Renee, but she still tended to use it against her. Theirs was certainly a toxic and troubled relationship, and, as Schwarzenbach admits, part of her drive to constantly depart had to do with attempts to flee her mother's stifling and damaging influence. The lack of positive female role models led her to look for female and motherly affection elsewhere, and this was what partly drew her to Maillart. She may also have had a bit of a crush on Maillart, but perhaps more than any actual physical intimacy she sought out her warmth and empathy.

Schwarzenbach's life was short and full of drama – not infrequently self-induced – ridden by personal struggles and marked by family trauma. Despite all that – or perhaps because of that, at least to a degree – she never surrendered in her intellectual, mental and physical efforts to be a vocal witness to injustice, evil, oppression and also beauty, which sometimes triumphed against the odds. Aware of her own limitations and vices, she persevered in her attempts to see good and hope in the world that she traversed so tirelessly across all kinds of division lines – geographical, political, religious and cultural alike. In her struggles, she more often than not failed, at times spectacularly, succumbing to her demons – drugs, unrequited affection, family feuds, physical and mental ailments, the most tragic and ironic being the 1942 biking accident near her home in Switzerland that sent her into a coma she did not recover from. When she died at the age of 34, Schwarzenbach was a half-reformed addict, a published author, an acclaimed journalist and social reformer, a married woman, a seasoned traveller and experienced driver, and – to quote her friend, the photographer Marianne Breslauer – 'this strange mixture of man and woman . . . not at all like a living being but like a work of art . . . the Archangel Gabriel standing before Heaven' (Breslauer 2012: 132). Breslauer's comment captures very well the conflicted nature of Annemarie Schwarzenbach and the ambiguity she exuded, eluding easy categorizations and definitions, which surfaces acutely in the countless pictures that Breslauer took of her. Breslauer was in fact not the only person to compare Schwarzenbach to an angel – so did Thomas Mann and Roger Martin du Gard; Carson McCullers recalled that Schwarzenbach's face 'would haunt [her] for the rest of her life' (Shepland 2021: 36). In the photographs Breslauer took of her, Schwarzenbach rarely smiles; rather, her usual expression seems to be defiance mixed with vulnerability, underrun by passion for whatever it is that the picture catches her doing. Her energy vibrates but the tremors sometimes send her over the edge.

Such was her condition in 1939 when she and Ella Maillart started planning the Afghan trip. Schwarzenbach, after a prolonged stay at a rehabilitation centre in Yverdon, appeared to have recovered just enough to embark on a new escapade. She knew she had nothing to lose – she had already lost nearly everything, several times at least – and she realized that there was actually quite a lot at stake for her, including the hope of learning to take control of her life, which she had long felt slipping away from her. Understanding that on her own she'd most probably give in to morphine addiction again, she found the level-headed and compassionate Maillart a perfect companion. Maillart, however, was not immediately convinced, especially given that some of her friends had warned her about Schwarzenbach. Though she tried not to let gossip get to her, she recalls towards the beginning of *The Cruel Way* that she 'was not quite easy about our enterprise' (Maillart 2013: 9). Schwarzenbach had her misgivings, too, but they were more about her general 'malaise' and the sense of impending doom she felt encroaching upon her:

> I am thirty. It is the last chance to mend my ways, to take myself in hand. This journey is not going to be a sky-larking escapade as if we were twenty – and that is impossible, with the fear of Hitler increasing day by day around us. The journey must be a means towards our end. We can help each other to become conscious, responsible persons. My blind way of life has grown unbearable. What is the reason, the meaning of the chaos that undermines people and nations? And there must be something that I am to do with my life, there must be some purpose for which I could gladly die or live. (Maillart 2013: 4)

Maillart could not help but be moved by such an ardent plea, to which she responded with a no less passionate 'prayer' for Schwarzenbach (whom she calls 'Christina' in the text in order to avoid raising objections from Schwarzenbach's family):

> May it be in my power to help you, impatient Christina so irked by the limitations of the human condition, so oppressed by the falsity of life, by the parody of love around us. If we travel together, may it be given me not to fail you, may my shoulder be firm enough for you to lean on. Along the surface of the earth I shall find our way where I have journeyed before; and inwardly, where I have long ago begun to ask myself questions so like yours, may the little that I have found help you to find what each of us has to find by himself. (Maillart 2013: 4)

For Jessa Crispin (also a travel writer and a journalist; author of the memoir-travelogue *Dead Ladies Project*), who wrote the most recent introduction (2013) to Maillart's *The Cruel Way*, it is the bond between the two women that occupies centre-stage in Maillart's account. Indeed, as one gets further and deeper into the narrative, one cannot help noticing that the entire text – and actually the entire endeavour – is completely about Schwarzenbach. Maillart does not hide it; quite the opposite, she states bluntly that her main goals 'were to acquire self-mastery and to save my friend from herself' (Maillart 2013: 26). Reading the two accounts side by side, the striking way in which they function might seem dubious at first because whereas Schwarzenbach's *All the Roads Are Open* offers a full-on, painstaking vivisection of its author, *The Cruel Way* provides very little in terms of a similar insight into Maillart's sensibilities and mindscapes. Rather, what is being revealed are the 'bumps' along the road that Schwarzenbach tended to conceal or at least downplay, and which Maillart supplied with background and context, laying bare the true emotional and physical cost of their journey.

One way to look at their complex relationship might be by approaching the two as travellers of the 'stereoscopic' order defined by Charles Forsdick in his analysis of, interestingly, a different pair of travel companions in which Maillart again was one of the counterparts (Forsdick 2009: 293). In an exploration of the textualizations of Maillart's and Peter Fleming's accounts of their 1935 journey to China, Forsdick draws attention to several issues that prove significant also in the case of Maillart's and Schwarzenbach's parallel travel narratives – the self-performativity of the travel writer, the narrative representation (or lack thereof) of the co-traveller, and the 'anxiety of influence' (ibid.: 295) that each of the travellers and authors may be subject to. Focusing on a journey undertaken by the 'odd couple', as Maillart and Fleming had been dubbed by Maureen Mulligan (2008: 141), Forsdick also looks at the gendered aspects of the travelling experience and the gender rivalry that also plays a part in how the authors' travelogues were textualized, noting that indeed each seemed to feel compelled to challenge the other in particular with regards to the 'ownership' of the journey (Forsdick 2009: 293) along the lines of the gender divide. The gender rivalry argument does not really hold for Schwarzenbach and Maillart. Even if Schwarzenbach's sexual and gender identification were not normative, they were, in essence, 'two women alone in Afghanistan' (Schwarzenbach 2018: 89) and had to fight the same battle, at least on this ground. However, questions of their authorial self-performativity, the presence of the co-traveller in each other's travelogues, and the anxiety of influence remain valid points.

Yet another example of the complexity and ambivalence inherent in the concept (and practice) of 'stereoscopic travellers' could be Graham and Barbara Greene's joint journey through Liberia in the early 1930s. In *Journey without Maps*, Graham Greene presents a vivid account of the trip, capturing the landscape, the people and the sociopolitical intricacies of the region. His narrative, however, tends to overshadow the contributions and perspectives of his cousin Barbara, who accompanied him and whose insights could have offered a more nuanced understanding of the culture they encountered. This dynamic raises questions about narrative authority and the implications of authorship in travel literature. Barbara Greene's perspective, although less represented, is essential for a holistic understanding of their travels because she provides an alternative view that contrasts with her cousin's often cynical outlook. Graham Greene's minimization of Barbara's presence in *Journey without Maps*, while reflecting his personal narrative style, acutely highlights the challenges faced by female voices in a predominantly male literary tradition. Still, the duality of their experiences aligns with the idea of stereoscopic vision, where the combination of two viewpoints creates a richer tapestry of understanding. In 1938 Barbara Greene published her own memoir of the trip, *Land Benighted* – republished in 1990 as *Too Late to Turn Back*. Critics' and readers' opinions remained divided on the question of how well the two accounts actually match up.

Maillart and Schwarzenbach's case, perhaps owing to the status of their relationship and their individual predispositions, is, similarly, not all too straightforward. Schwarzenbach surely takes great care of what she puts into her writing, diligently avoiding mentioning any of the crises caused by her recklessness, which may be read as an attempt at preserving the image and impression of a sturdy traveller and a rational and responsible person, but then again she does not hide her depressive moods, ennui, fear and insecurity, at times allowing herself to wallow in self-pity and painstakingly record her anxieties. Reading Schwarzenbach's *All the Roads Are Open* leaves one in the dark with regards to what went on 'behind the scenes'; her mood swings are recorded but Schwarzenbach is rather cryptic when it comes to revealing the full picture. This in turn seems to suggest that although she was publishing her journey reports regularly while on the road, she was very careful about how much she actually revealed, signalling a very conscious and self-aware authorial persona.

Perhaps she kept a more intimate diary or shared the many dramas of the journey in letters to friends; sadly, these would not have survived her mother's censorial hand. Surely the travel experience was shaping her travelogue to a great extent, which had to do with the physical aspect of her engagement in the entire endeavour. What is quite striking, though, is that Schwarzenbach erases nearly all traces of the presence of Maillart, going so far to as to refer to her rather impersonally as 'my companion' or 'my fellow driver' in the rare instances when she mentions her at all (Schwarzenbach 2018: *passim*). Forsdick suggests that similar practices in Maillart's and Fleming's respective Chinese travelogues exemplified a Romanticist apotheosis of solitary travel (Forsdick 2009: 295) and the cult of the solo traveller who, consequently, becomes a true master of all they survey. Regarding Schwarzenbach's and Maillart's Afghan journey this does not seem to be the case, and the grounds for Schwarzenbach's textual decisions might actually lie elsewhere, in a realm altogether much more personal and symptomatic of her condition. To put it perhaps a bit too bluntly, Schwarzenbach might not have been particularly interested in such implications of the journey because she was far too engrossed in her own universe, to the point of sometimes seeming not to be of this world. Focused inwardly, she appears throughout to be devoted to a self-vivisection that does not require her to always consider the bigger picture. Interestingly, while Schwarzenbach erased the travel/travail parallel in her account, Maillart embraced and highlighted it. Thus, it might be argued that Schwarzenbach's *All the Roads Are Open* constitutes a first-person, first-hand account of a heroine's journey – albeit a selective one – while Maillart's *The Cruel Way* becomes a companion piece to the record of Schwarzenbach's quest.

Indeed, their Afghan trip possesses a number of features typical of the hero's quest, both with regards to the reasons that they set off in the first place, and in terms of the particular stages that their journey is made up of. Here, Maureen Murdock's work comes in especially handy – it is she who built, or expanded, the framework of the heroine's journey, extricating it from Campbell's heroic quest. As Murdock puts it:

> The journey begins with our heroine's search for identity. This 'call' is heard at no specific age but occurs when the 'old self' no longer fits . . . It may simply occur when a woman realizes that she has no sense of self that she can call her own. (Murdock 2020: 5)

Usually, the sense of lack, confusion, stasis and pointlessness goes as far back as the archetype of the mother and the heroine's urge to separate from it in order to find a truer self that would be fuller and individual. This entails both the separation of a woman from her actual mother, and the deeper, more complex split required of her at the level of the unconscious, involving body and mind.

According to Murdock, 'geographical separation may be the only way at first to resolve the tension between a daughter's need to grow up and her desire to please her mother' (Murdock 2020: 20). Having managed to acquire some safe distance, the heroine may brace herself to tackle the archetypal mother and her dual expressions – that of the good Great Mother, and that of the Terrible Mother, who embodies stasis, stagnation and whose hold on the daughter petrifies her, ultimately threatening her with death. Renee's actions are frequently hostile, but Schwarzenbach, though at least half-aware of their detrimental effect on her health, self-esteem and well-being, finds herself helpless and all-too-exposed to their impact. She confides in Maillart that she thought 'it was very foolish of me to be always acting against my mother – the person who knows me better than anyone else. I had no hope of freeing myself from her, no hope of ever being simply myself' (Maillart 2013: 15). Maillart, while sensing Schwarzenbach's fear of Renee, noticed as well that Schwarzenbach might have been just as ready as one can get to break the circle and split herself from her mother, in an act of defiance akin to taking the sword of her truth and using it to protect herself along the path of her own destiny. The first step has been made – Schwarzenbach takes to the wheel of their Ford; but does she have enough stamina and courage to keep at it? Alone with her baggage, probably not; but with an ally at her side the chances of success grow. Thus, in an attempt – as noted earlier – to 'save Christina from herself', Ella Maillart accepts the role of the guardian and nourisher, providing Schwarzenbach with models of femininity that are powerful but not threatening, unfurling before her a path that, while still full of traps, allows her to learn to use her inner strength to manoeuvre around the pitfalls.

## The Fear of Immobility

Schwarzenbach's and Maillart's road remains one of trials, with calamities and losses on the way. The second most dangerous 'ogre', after the fear of the mother, was always Schwarzenbach's fragile sobriety and the temptations offered by places they passed by, especially cities such as Budapest, Sofia and Constantinople, or indeed their final destination, Kabul. In Sofia, Maillart was at first totally at a loss with regards to what might have happened; when she discovered a broken ampule in the bathroom she quickly understood what was going on:

> She had succumbed once more ... She had done what she pretended to abhor. My presence, my confidence in her, the fear of displeasing me had no effect. What did it mean? What were we to do? (Maillart 2013: 24)

The transgression leaves Schwarzenbach seriously incapacitated – a fact we learn however from Maillart because Schwarzenbach's own account includes only a couple of cursory remarks on Bulgaria, quite evasive in tone and devoted to landscapes, roads and availability of amenities such as fresh bread (Schwarzenbach 2018: 4–7). It is Maillart who reveals the gravity of the situation, but fortunately she was not one to break down and lose her head in a crisis; even though at this particular moment she felt betrayed and dismayed, her fears for her companion's health and life prevailed and she set about making sure this would not happen again. To this end, she and Schwarzenbach made a pact, an unwritten but binding contract according to which Schwarzenbach gave herself over to Maillart: '"I give you complete power day and night. Don't leave me alone. If it happens again, I leave the car with you and go back"' (Maillart 2013: 34). Profound though these pronouncements were, at the end of the day they were only words, and unfortunately, Schwarzenbach gave in to temptation yet again; but this time her condition was even graver. The second crisis hit towards the end of their journey, when they were already in Kabul. At first, she seemed to be coming down with a bad flu, but despite a high fever and severe cough she refused to see a doctor. As Schwarzenbach's state worsened, Maillart found herself feeling more and more suspicious. Maillart noted with alarm that 'she was ill but would not admit it; she would not relax or submit herself to the laws of nature' (Maillart 2013: 197); what she found yet more unsettling was the gnawing anxiety that she herself had let Schwarzenbach down:

> I failed Christina. According to our pact I was not to leave her alone whatever she might say or do. But the intensity of my desire to help her had spoilt my intension. That intensity had brought with it a kind of effort that had tired me. (Maillart 2013: 198)

When Schwarzenbach finally speaks, it is 'a confession of her total wretchedness' (Maillart 2013: 199). She admits to having been lying all the time, procuring morphine ampules whenever they were in the vicinity of any bigger town. Maillart is distressed by the confession, yet somehow it does not feel entirely convincing to the reader that she has had no inkling whatsoever of what Schwarzenbach had been up to, especially as she forgives once again this blatant breach of their contract, which jeopardized not just the journey but their friendship and Schwarzenbach's life.

Despite Maillart's disappointment, they trudged on, though Schwarzenbach was fragile and mentally and physically weak. Neither was she the easiest of convalescents – she vehemently refused to see a doctor and at times seemed to take a peculiar tormented pleasure in wallowing in her afflictions. Maillart, who saw through all this, began to lose patience, only to backtrack on any nascent criticism of her friend, starting instead to look for explanations and reasons that Schwarzenbach was this way. The level of her compassion is astounding; it is also at this moment, after Schwarzenbach has worn out nearly all of Maillart's goodwill and care, that Maillart utters what may very well be one of the most moving and saddest remarks on her companion: 'She was a sober violin whose cords had been made of her own heart-strings: while she played she was wearing herself out' (Maillart 2013: 73).

No reflection of these crises finds expression in Schwarzenbach's account, but one may detect her fear and internal turmoil in the violent mood swings she succumbs to. Describing one of the villages they stop by, she resorts to a language of loss, defeat and death; the settlement comes across to her as 'gripped by the inexorable destruction like a contagious disease' (Schwarzenbach 2018: 65); all around she sees ghosts and ruins of things that have once been and are no more:

> And I look about, and nothing remains, the bells are buried, the cisterns caved in, the altars cold. In this merciless land, one is tempted to believe the earth is on the verge of extinction. (Schwarzenbach 2018: 66)

A moment later, though, she lets herself be gripped by the nearly all-consuming beauty of their surroundings and the kindness of the people they encounter, their humility and respect for the timeless rites of their land:

> And when you see a white or a light blue turban, a women's red veil by day in the fields, or at noon the men who have gathered to pray outside the little mosque, or in the evening, at the entrance to the bazaar, the first warm glow of a samovar, you believe the course of a well-ordered day is assured, a well-ordered life according to laws devised for our protection, for our needs, and you feel almost sheltered. (Schwarzenbach 2018: 70)

Another day on the road brings a new storm, an unease and listlessness from which she 'wanted to break away, not knowing exactly from which fate, seeming to grasp only that I had been struck by calamity' (Schwarzenbach 2018: 100–1). Not only does she feel drained mentally; she ponders withdrawing from professional activity, too – writing becomes 'a perpetual mirror of our drenched existence, which I was so loath to accept and endure' (ibid.: 101). The torments of travelling and writing draining the author of the *élan vital* necessary to keep going, both text- and distance-wise, were perhaps best expounded upon by T.E. Lawrence – though despite the masochistic urges he so often succumbed to, he managed to write *Seven Pillars of Wisdom* (1926) before giving in to the death drive. Describing his writing method, Lawrence says: 'I tie myself into knots trying to re-enact everything, as I write it out. It's like writing in front of the looking-glass' (Lawrence quoted in Białas 2006: 73). The condition of being 'tied into knots' is surely an uncomfortable one; when applied to the process of textualizing one's narrative account of the trials and tribulations of travel, it becomes even more symptomatic of how the author's inclinations might exhaust him. Of course, Lawrence makes it yet more problematic because he relishes any prospect of self-victimization and suffering, which to an extent might be said of Annemarie Schwarzenbach, too. After all, she also sought pain and frequently gave in to self-destructive impulses. The resemblance however seems to end here: Lawrence pursued sexual gratification and was masochistically attracted to the degradation of the body and spirit that brought him release and satiation; Schwarzenbach's masochistic tendencies were always about the annihilation of the ego, which she saw her as greatest nemesis.

From a different angle, Lawrence's accounts of travels in the Middle East during the First World War do share some affinity with Schwarzenbach's narrative, fraught with the tensions of the encroaching Second World War. Both authors moved and wrote under liminal circumstances, painfully aware of the potential consequences of war in a global, as well as personal, dimension. However, the causes of Schwarzenbach's spleen in this instance perhaps had to do with having stayed in Afghanistan too long, allowing herself to soak in the place and its circumstances – an occasion that always troubled her because it seemed to foreshadow growing roots, committing to a place and to a people, and in effect becoming 'immobile' and giving in to the stasis that she so feared. Maillart, now already an expert at diagnosing Schwarzenbach's states, noted the mood swings with alarm:

> One day working hard, announcing joyfully that she was rid of her obsessive fear; the next confessing that she could not live much longer in such dismal unhappiness – it was obvious that she was not nearing a well-balanced state. (Maillart 2013: 85)

Later, Maillart confirms what Schwarzenbach has revealed in her outbursts – that 'she could not bear the thought of remaining inactive, for then she felt like dying. She was afraid of immobility' (Maillart 2013: 197). The terror Schwarzenbach experienced whenever so much as the prospect of inertia presented itself speaks to a malaise shared by many modernist authors, artists and intellectuals alike who, as noted by Andrew Kalaidjian, exhibited 'a perpetual investment in new ruptures and revolutionary potentials' (2014: 38) in the belief that 'the wholesale rejection of inertia [would lead] to artistic renewal' (ibid.: 39). In Annemarie Schwarzenbach's case, immobility and inertia were trigger warnings that launched her into a self-redefinition that she was bent on executing regardless of the costs it would lead her to incur.

## Conclusion: Terror, Pity, Love

That Maillart endured all the disappointments and betrayals speaks volumes about her as a person of deep compassion and empathy. However, given that *The Cruel Way*, in contrast to *All the Roads Are Open*, was published in 1947 – nearly eight years after the trip and five after Schwarzenbach's death – Maillart's pensive mood might also signify her wistfulness at the course of events. Schwarzenbach was publishing her travel impressions and reports during the trip, submitting them to various magazines and journals without allowing herself much time to revise and edit; Maillart worked on her travelogue on and off until its publication and surely benefited from hindsight, at least with regards to how she viewed Annemarie and her struggle. Schwarzenbach's biking accident in Sils in 1942, coma and ensuing death shook Maillart deeply, which may have had an impact on how she looked at her friend and what she actually put into her work. She certainly arrived at an empathetic understanding of Schwarzenbach, the kind that neither Claude Clarac – her partner in an ill-suited marriage of convenience – nor her mother ever achieved. She saw Schwarzenbach's death drive, her pursuit of suffering and her drug abuse as tragic attempts at what Schwarzenbach called 'lessening the small ego' (Maillart 2013: 205) – overcoming suffering through confronting it head-on, seeking it out so as to defeat it, or at least numb it enough to make it bearable. In this, perhaps Schwarzenbach found deliverance; if not, then surely not damnation, either.

If escape – total 'disembarkation' – was what Annemarie Schwarzenbach was after, travel (and travel writing) proved to be a fitting practice to engage in. According to Karen Lawrence, 'travel writing reveals a set of alternative myths or models for women's space in society – against the myths of dependency, women's inferiority, and romantic love' (Lawrence 1994: xi), one that is achieved – and achievable – only when women shake off the preassigned roles of 'beloveds' or 'homemakers'. That, however, does not have to mean a straightforward 'either/or' because in their performance of the parts of travellers and storytellers, women can embody this seemingly exclusive duality. Through what Jacques Derrida calls 'breaking the law of boundaries', a woman on the move is at once 'a weaver [and an] unweaver, a constructor [and a] deconstructor' (Derrida 1982: 70). This new realm has been cleared out for women specifically by the options that travel writing opens up through providing 'discursive space for women who sometimes left home in order to write home, discovering new aesthetic as well as social possibilities', as Lawrence (1994: 18) puts it. Lawrence has an important point here, one that Annemarie Schwarzenbach and Ella Maillart prove in their respective works: both women have managed, in their unique ways, to appropriate and shape the travel genre to make it fit their own experience and answer to their own goals, thus creating what Lawrence defines as a 'permeable membrane of possibilities' granted by the unknown (ibid.), be it a concrete geographical territory, a more 'ephemeral' realm or an inner landscape (or, indeed, all of these at once). Each of them seized on the chance to approach that which had been repressed: in Schwarzenbach's case, the archetypal mother and her physical emanation in the figure of Renee; for Maillart, the long-unvoiced but yearned-for taming of the monstrous, constantly craving ego. Naturally, success in acquiring equilibrium cannot be guaranteed and thus should not be taken for granted; many traps lurk along the way and threaten to send the seeker, the traveller, back to square one, as it were, or worse to have her choke on the desired object and give in to melancholia, neurosis or irreconcilable grief. Still, the risk is (probably) worth taking.

Annemarie Schwarzenbach was never one to avoid risk; quite the contrary, she recklessly pursued it and invited it to lead her down many a shadow-cast valley. She hardly ever celebrated outright success in her endeavours, more often than not falling prey to the perils with which the paths she chose had been strewn. Nonetheless, again and again she rose, no matter how deep the bottom she had hit. Having come precariously close to death on the Afghan trip with Maillart, she rebounded and came to a new reckoning with herself that she eagerly shared with Maillart, proclaiming with confidence: 'I know now, smoke or dope are useless ... the drug ... was the fatal desire to kill life, to wipe out pain and joy, the tension-source of human activity' (Maillart 2013: 206). Maillart cheered for her, albeit cautiously, but appeared to dare to believe this was it, the sought-for boon of success – saving Schwarzenbach from herself:

> Her eye was once more filled with light, her body poised, the past not weighing on her. She was thinner than ever but healthy ... The past had lost its bitterness. (Maillart 2013: 202)

Maillart closes her account here, leaving Schwarzenbach with her new-found balance and peace of mind. However, writing in 1947, she carried the full burden of hindsight over the later vicissitudes of Schwarzenbach's fate. The irony could not be bitterer. In a song she wrote about Schwarzenbach, Suzanne Vega aptly remarks: 'Annemarie ... terror, pity, love' (Vega 2011), and the order needs no reconfiguring – pity, yes; terror – constantly; love – hopelessly, recklessly, always.

And Annemarie Schwarzenbach herself? She concludes *All the Roads Are Open* with a poignant observation that 'what staggers us, over and over again, is the morning splendor of departure' (Schwarzenbach 2018: 120). Touché.

An earlier version of this chapter appeared in *Crossroads: A Journal of English Studies* 39 (2022) under the title 'How to Disembark Completely: Annemarie Schwarzenbach's and Ella Maillart's 1939 Afghan Journey'. The initial drafting and editorial input for this previous version, however, were undertaken as part of the early stages of the *Writing on the Move* project.

**Dr Julia Szołtysek** is Assistant Professor at the University of Silesia in Katowice, Poland. Her academic interests include literary and artistic representations of the Middle East, travel writing, contemporary British and American fiction by women, and opera studies. She is the recipient of the 2016 Peter Lang Young Scholars Award. Her monograph *A Mosaic of Misunderstanding: Occident, Orient, and Facets of Mutual Mis/Construal* was published in 2016 by Peter Lang.

## References

Białas, Zbigniew. 2006. *The Body Wall: Somatics of Travelling and Discursive Practices*. Frankfurt: Peter Lang.

Boswell, James. 2004 (1785). *The Journal of a Tour to the Hebrides with Samuel Johnson, LL.D.* E-book. Project Gutenberg.

Breslauer, Marianne. (2009) 2012. *Bilder meines Lebens: Erinnerungen*. Wädenswil: Nimbus.

Campbell, Joseph. 2004 (1949). *The Hero with a Thousand Faces*. Princeton, NJ: Princeton University Press.

Conway, James J. 2012. 'Death in Persia', *Strange Flowers*, 15 November. Retrieved 2 August 2024 from https://strangeflowers.wordpress.com/2012/11/15/death-in-persia/.

Crispin, Jessa. 2015. *The Dead Ladies Project: Exiles, Expats, and Ex-Countries*. Chicago: University of Chicago Press.

Decock, Sofie. 2011. '"The Loving Conquest and Embrace": On Peaceful Heterotopias and Utopias in Annemarie Schwarzenbach's Asian and African Travel Writings', *Women in German Yearbook: Feminist Studies in German Literature & Culture* 27: 109–30.

Decock, Sofie, and Uta Schaffers. 2011. '"Als ob ich dort noch etwas zu suchen haette": Europa- und Orient-Diskurse in Annemarie Schwarzenbachs Erzahlung', *arcadia* 46(2): 406–24.

———. 2012. 'Die Suche nach bergenden Raumen als Reise und Schreibprogramm: Annemarie Schwarzenbachs journalistische Afrika-Texte', *Orbis Litterarum* 67(1): 1–24.

———. 2018. 'Die Darstellung der "anderen Frau" in Reisetexten: Eine diskurslinguistische Analyse', *Literaturwiss Linguistik* 48: 775–97.

Derrida, Jacques. 1982. 'Choreographies', *Diacritics* 12: 66–76.

Dubini, Donatello, and Fosco Dubini (dirs.). 2001. *The Journey to Kafiristan*. Film. Real Fiction Filmverleih.

Farley, David. 2016. 'Modernist Travel Writing', in Carl Thompson (ed.), *The Routledge Companion to Travel Writing*. Abingdon: Routledge, pp. 278–87.

Forsdick, Charles. 2009. 'Peter Fleming and Ella Maillart in China: Travel Writing as Stereoscopic and Polygraphic Form', *Studies in Travel Writing* 13(4): 293–303.

Greene, Barbara. 1990 (1938). *Too Late to Turn Back*. London: Penguin.

Greene, Graham. 1988 (1936). *Journey without Maps*. London: Penguin.

Kalaidjian, Andrew. 2014. 'Positive Inertia: D.H. Lawrence and the Aesthetics of Generation', *Journal of Modern Literature* 38(1): 38–55. https://doi.org/10.2979/jmodelite.38.1.38.

Lawrence, Karen R. 1994. *Penelope Voyages: Women and Travel in the British Literary Tradition*. Ithaca, NY: Cornell University Press.

Maillart, Ella. 2013 (1947). *The Cruel Way: Switzerland to Afghanistan in a Ford, 1939*. Chicago: Chicago University Press.

Maruo-Schröder, Nicole, Sarah Schäfer-Althaus and Uta Schaffers. 2023. *Traveling Bodies: Interdisciplinary Perspectives on Traveling as an Embodied Practice*. New York: Routledge.

Mazzucco, Melania. 2000. *Lei cosi amata*. Milan: BUR.

———. 2006. *Tak ukochana*, trans. M. Woźniak. Warsaw: Wydawnictwo WAB.

Mulligan, Maureen. 2008. '"The Controversy Still Rages": Ella Maillart and Peter Fleming: Forbidden Journeys to China and Beyond with the Odd Couple', in Steve Clark and Paul Smethurst (eds), *Asian Crossings: Travel Writing on China, Japan and South East Asia*. Hong Kong: Hong Kong University Press, pp. 141–48.

Murdock, Maureen. 2020 (1990). *The Heroine's Journey: Woman's Quest for Wholeness*. Boulder, CO: Shambala.

Schwarzenbach, Annemarie. 2011. *Lyric Novella*, trans. L. Renner Jones. London: Seagull.

———. 2012. *Death in Persia*, trans. L. Renner Jones. London: Seagull.

———. 2018 (2011). *All the Roads Are Open: The Afghan Journey*, trans. I. Fargo Cole. London: Seagull.

Shepland, Jenn. 2021. *My Autobiography of Carson McCullers*. London: Virago.

The Ravaged Angel. 2021. 'On a Journey the Face of Reality Changes'. Tumblr blog post. Retrieved 27 June 2021 from https://the-ravaged-angel.tumblr.com/post/618929375872712704/on-a-journey-the-face-of-reality-changes-with-the.

Thompson, Carl (ed.). 2016. *The Routledge Companion to Travel Writing*. Abingdon: Routledge.

Vega, Suzanne. 2011. 'Song of Annemarie'. Song from the stage play *Carson McCullers Talks about Love*.

CHAPTER 5

# The Novel at the End of the Road

## *The Soul Circuit of Jack Kerouac as a Professional Writer*

Martin Wable

### Introduction

The American writer of Canadian French-speaking descent Jack Kerouac (1922–69) implemented a topographical imaginary that redefined the literary canons of his century. Both travel writing and non-fiction literature in general owe much to the unbridled prose of his story *On the Road* (first published in New York in 1957, by Viking; all citations in this chapter are from the 2011 Penguin Essentials edition). While the road is fundamentally linked to motion, the road also reaches its end, like any novel. Whether the end of the road is simply construed as the destination (the end of a country) or as a potential redemption (the end of oneself), it is full of promises and disappointments. In Kerouac's work, the breathless swerves of the narrator, of his friend Dean and of all the other characters who mingle with their movements, in the spirit of improvisation, are orchestrated in the repetition of the 'end of' pattern and its main variations. This recurring motif reflects a dynamic between 'on the road' and what we might call 'off the road'; in other words, between the times when Kerouac, under the pseudonym of Sal Paradise, is in motion, and the times when he is stationary or sedentary. This can be seen in the way Sal Paradise sometimes observes a disenchanted cityscape from a cheap hotel room, which serves as his home base after long journeys, such as here in Los Angeles:

> We were very happy in our little hotel room. In the middle of the night, I got up because I couldn't sleep, pulled the cover over baby's bare brown shoulder, and examined the LA night. What brutal, hot, siren-whining nights they are! Right across the street there was trouble. An old rickety rundown rooming house was the scene of some kind of tragedy. The cruiser was pulled up below and the cops were questioning an old man with gray hair. Sobbings came from within. I

> could hear everything, together with the hum of my hotel neon. I never felt sadder in my life. LA is the loneliest and most brutal of American cities; New York gets god-awful cold in the winter but there's a feeling of wacky comradeship somewhere in some streets. LA is a jungle. (Kerouac 2011: 77)

Here we see how the moments when Sal Paradise stops on the road to settle down for a few days lead him into a terrible impasse and a longing to run away again. It is this same interest in the end of oneself that pushed Kerouac, during his career as an author, to recognize himself beaten, dejected, at the end of his journey, dented by existence, giving a new meaning to the term 'beat', sublimated in the idea of the Beat Generation, in which Kerouac finally found a new community at the other end of himself and a disenchanted America.

The use of computer language processing tools (NLP, natural language processing) has enabled us to extract passages of text containing the 'end of' pattern. Kerouac uses 'end of the road' several times and also uses expressions such as 'end of the continent/end of America'. The aim of this NLP exploration was to see when these phrasings occur and how often. The first occurrence of 'end of the road' appears quite late in the text, on page 202.[1] Its use then accelerates, on pages 209 and 222. Then we find the expression at the end, on pages 251, 274 and 275. On the other hand, 'end of America' first appears earlier, on pages 70 and 74 (with the last occurrence on page 249). And 'end of the continent' also appears before the first occurrence of 'end of the road', on pages 122, 161 and 183. We note a total of twelve occurrences of the motif under its forms 'end of the road/the continent/America'.[2] The general tendency is for this pattern to be repeated at the end of the story. In addition, we have identified a series of forms that appear only once with the phrase 'end of': end of a doubt, end of the line, end of the country, end of each little street, end of a town, end of a field, end of a row, end of something and so on.[3]

This analysis leads to the conclusion that Kerouac is above all a professional writer before being a traveller lost on the paths of this world. He uses all the materials of wandering, of his personal experience and of his theories or literary intuitions of his time. His doubts about his future as a writer are the doubts that any young person might have about their future career.[4] Moreover, Kerouac is a writer with a vision of the road and the 'end of the road' encapsulates his broader literary approach as a writer. In the novel, unlike his friend Dean Moriarty, Sal Paradise is scattered, but finds himself, as he uses the material of 'the road' to make his novel. In real life it was not without a feeling of imposture and guilt, about his authenticity as a traveller, a Buddhist or Christian pilgrim, that Kerouac faced his success in the last decade of his life.[5] Indeed, he blamed himself for having made his friends popular characters, and he felt unable to assume his role as king of the beatniks, which the American media gave him, when his book *On the Road* met, six or seven years after its initial writing, with unexpected success. Kerouac's chance to finally be able to 'find himself' at the end of the road is, therefore, the redemption the writer was hoping for. This is also what makes Kerouac's work a novel of the soul.[6] The road is an invitation for everyone, whatever the conditions of a country or an era, to try to get closer to their soul.

In order to analyse the text as an itinerary and explore how the meaning and stakes of the 'end of' motif evolve, this chapter is split up into four sections, each corresponding to a trip the narrator takes. The first one leads to the end of a subjective story-to-tell. We then explore the depths of America and the Beat consciousness. The 'end of' motif then shifts towards the expression of the absurdity of the mid-twentieth-century human condition and to the human comedy. The last section deals with redemption at the end of the story.

## The First Trip, Ends of the Land, Ends of the Subjective Story-to-Tell

A chronological analysis of the motif of the end of the road allows us to explore the evolution of its impact on the poetics of movement and stasis. The first time Kerouac uses this motif is in a context where the character he portrays, Sal Paradise, arrives in Chicago. After a day on the road, and a day spent in a cheap hotel room, he walks to discover the city when the night comes. Sal evokes the recent history of jazz bop, in this era when the great masters follow one another, from Charlie Parker to Miles Davis, and it seems to him that this whole musical revolution is in the air of the city. He sits down to listen to 'the sound of the night', and then he thinks of his friends across the United States:

> And as I sat there listening to that sound of the night which bop has come to represent for all of us, I thought of all my friends from *one end of the country to the other* [emphasis added] and how they were really all in the same vast backyard doing something so frantic and rushing-about. And for the first time in my life, the following afternoon, I went into the West. (Kerouac 2011: 13)

The motif that here represents the end of the country can be read in a geographical sense, although it is a geography that is more fantastic than real. In a lyrical tone, through the voice of the narrator Sal Paradise, Kerouac imagines a vast America that ends in the East and the West. In the middle of this expanse, which seems even more infinite in the night of his solitary wandering, he remembers his friends scattered from one cardinal point to another. The author invites the reader to become aware of the almost non-human geographical space in which friendship takes place, which is based on emotional values. This image inspires Sal to move to the West for the first time.

We also find the motif much later in the novel, in the twelfth chapter. Kerouac uses it with great emphasis on the dramatic nature of the situation. His fictional counterpart, Sal Paradise, is not only at the end of the road, and at the end of the continent, since he is now at San Francisco, but also at the end of himself. The motif appears during Sal's meeting with the father of his friend Remi Boncœur, which turns out rather badly. Remi has asked him to behave well for his father's visit to San Francisco, but Sal arrives drunk and late, and the embarrassment builds up. The narrator concludes that he can no longer draw anything positive from his trip: 'Here I was at *the end of America* [emphasis added] – no more land – and now there was nowhere to go but back. I determined at least to make my trip a circular one: I decided then and there to go to Hollywood and back through Texas to see my bayou gang; then the rest be damned' (Kerouac 2011: 70).

Sal is nearing 'the end' of this side of America, the western border where the Pacific Ocean lies. He is in a geographical impasse that is also personal, since he can no longer stay with his friend Remi Boncœur. He must make a decision and this decision translates for him into the choice of the next destination. His plan is not to return to New York, but to make a loop. He initially considers Texas, but eventually, unexpected events will derail his plans and introduce another motivation for his journey. He meets Terry, and a weeks-long romance unfolds. Above all, this story is an adventure in which the narrator questions himself. As Sal and Terry arrive in Los Angeles by coach, Sal looks out the window, and the idea of the end of America emerges in his daydreams:

> I looked greedily out the window: stucco houses and palms and drive-ins, the whole mad thing, the ragged promised land, *the fantastic end of America* [emphasis added]. We got off the bus at Main Street, which was no different from where you get off a bus in Kansas City or Chicago or Boston – red brick, dirty, characters drifting by, trolleys grating in the hopeless dawn, the whorey smell of a big city. (Kerouac 2011: 74)

These reveries are bittersweet and are characterized by the observation of the margins at the end of the continent. Yet Kerouac finds this end of America bordering on the dreamlike 'fantastic' interpretation he has made of it.[7] Ann Charters has pointed out that he was inclined to see his own legend in the simplest of things, which can be understood as a tendency to hyper-narrativize one's own life: 'His talent as a writer was not his inventiveness with new characters and plots, but rather his power to dramatize the spirit of his own life into romantic fantasy' (1994: 63).

Moreover, in the preceding extracts, the end is no longer so much geographical as social. The end of America, in other words, is used by Kerouac to refer to the bottom of America: tasteless landscapes, poor people, an atmosphere of sadness and poverty. The dimension that Kerouac gives to the end of the road is part of this triptych that Claude Lévi-Strauss proposes as a definition of travel: 'We generally conceive of travel as a movement in space. That is not enough. A journey takes place simultaneously in space, in time, and in the social hierarchy' (2013: 92).[8]

In addition, to understand the implications of the motif under study, we need to consider how the character of Sal Paradise, as an individual in the 'too-big world',[9] can be portrayed as someone speaking about the human condition. Henri Meschonnic, in his critique of Theodor Adorno's work, emphasizes the difference between the individual and the subject – something that can cause ambiguity in biographical poetry or literature.

> It is a question of reducing the individual-confused-with-the-subject to the social. Because the subject is indeed social. And that the individual is supposed to be the anti-social, the incompatible – instead of the collectivity only existing if it exists. But a romanticism of the masses is substituted for it. Power strategy. Since we speak in his name: 'The work of art is social through the individual, without his being thereby aware of society: perhaps all the more so because he is less conscious.' [Adorno 1974: 222–23] (Meschonnic 2009: 94)[10]

This extract from Meschonnic highlights how Kerouac's solitude serves a social role in literature. Paradoxically, the public identifies more with him as an individual than with the 'normal' masses representing social order. This reflects the poet's power in society, reinforced by a romanticized individuation. Kerouac embraces this dynamic. In his portrayal of this 'off the road' moment, he reveals America's grey areas and the struggles of its people. By romanticizing loneliness and marginalization, he gives his voice a political purpose. Omar Swartz notes that 'the primary advantage of liminality in *On the Road* drama is its ability to shed light on the human condition', citing Victor Turner and David Hume to emphasize the concept of a marginalized outsider as symbolizing and embodying a universal human sentiment (Swartz 1999: 99). At the 'end of the continent' and 'end of America', Kerouac leads us into a dead end – one where neither the individual nor the character finds purpose. There is no story left, no hope of integration. But does the journey truly end? This narrative device keeps us waiting for Sal's return 'on the road', where the story belongs.

Sal becomes a cotton picker and completely forgets his trip and his friends: 'The days rolled by. I forgot all about the East and all about Dean and Carlo and the bloody road' (Kerouac 2011: 88). Here, again, we can identify the end 'of something'. It is not only a matter of geography. Despite all the conviction he may have felt in his adventure with Terry, Sal sees the end: 'I walked down the highway to Sabinal, eating black walnuts from the walnut tree. I went on the SP tracks and balanced along the rail. I passed a watertower and a factory. *This was the end of something* [emphasis added]. I went to the telegraph office of the railroad for my money order from New York' (ibid.: 92).

The end here is that of an episode into which the narrator, like the reader, has been drawn; but ultimately, it is the road that prevails to the detriment of the sedentary values represented by romantic connection, enrichment and work. The sudden change of context produced by the end of the adventure with Terry invites us to put into perspective the affective relationships in which human beings engage in the common spaces offered to them by American society in the mid-twentieth century. In terms of rhythm, this change of context follows the classic procedures of travel writing, notably the classic epic. The author adheres to an episodic narrative in which one event follows another. Homer's *Odyssey* comes to mind, in which Odysseus never stops for long during his adventures, whatever the charms that certain destinations offer him: wealth, pleasure, notoriety, stability.

Sal Paradise decides to return home. An additional version of our motif can be found in the posthumously published 'original scroll', while Sal Paradise is in Los Angeles, where he has to catch his bus back east:

> On the sidewalks characters swarmed. Everybody was looking at everybody else. It was *the end of the continent* [emphasis added], no more land. Somebody had tipped the American continent like a pinball machine and all the goofballs had come rolling to LA in the southwest corner. I cried for all of us. There was no end to the American sadness and the American madness. Someday we'll start laughing and roll on the ground when we realize how funny it's been. (Kerouac 2007: 206)

In the recurring alternation between the phrases 'end of the continent' and 'end of America', we can clearly establish the relationship between these two ideas. We also note that the same precision, 'no more land', was previously associated in the text with the other phrase ('[h]ere I was at the end of America – no more land'; Kerouac 2011: 70).

After travelling 8,000 kilometres across the continent, Sal Paradise arrives on the other side, the Atlantic side of America, 'the other end of the land'. This 'end of country' is characterized by high towers and crowds, as if the charm of travel, evoked in the first moments in Chicago, had faded. 'The high towers of the land – *the other end of the land* [emphasis added], the place where Paper America is born. I stood in a subway doorway, trying to get enough nerve to pick up a beautiful long butt, and every time I stooped great crowds rushed by' (Kerouac 2011: 96).

The narrator is confronted with a reality in New York that is still different from that of San Francisco and Los Angeles, but in which the search for meaning is associated with the idea of the end. The end here designates the eastern continental or Atlantic limit of the United States, the one quoted at the beginning of the book: 'I thought of all my friends from one end of the country to the other' (Kerouac 2011: 13). We are here at the end of the first part, and the first itinerary fictionalized by Kerouac is, we may say, complete. It is winter, the travel season seems to be over: 'the first cold winds rattled the windowpane, and I made it just in time' (ibid.: 97), the narrator reassures us on his return to his aunt's residence.

## The Second Trip, from the Depths of America to the Beat Revelation

The continent-ending motif returns at the start of Part 2, Chapter 6. This is Sal Paradise's second major continent-crossing. Despite everything, Sal fears finding himself alone once again at the end of the continent: 'The understanding had been that Marylou would switch to me in Frisco, but I now began to see they were going to stick and I was going to be left alone on my butt at *the other end of the continent* [emphasis added]' (Kerouac 2011: 122).

Once again, Kerouac invokes a longitudinal limit of the United States, this time the West Coast. The author insists on the geographical continuity of the continent, an object of imaginary investigation and literary research. This object also reflects the size of Kerouac's ambition, in line with other major American novelists who have investigated the idea of 'the American dream'. Marcel Labine writes of the Dean–Sal duo that 'they seek the secret of a once sacred America, that of the pioneers and founding fathers' (2002: 65),[11] thus inscribing Kerouac's novel in the genealogy of the American myth.

Sal's anxieties are well founded: after a few days spent at the house of their friend Old Bull Lee (whose character is based on William Burroughs) and having just arrived in San Francisco, Dean abandons Sal and Marylou on the sidewalk, to find Camille, his wife. Then Marylou abandons Sal too, leaving him alone and penniless in a cheap hotel. Here we find two other recurring motifs in Kerouac: the narrator-protagonist who looks out of his window, and the cheap bedroom – previously encountered during his night with Terry in Los Angeles (Kerouac 2011: 77). This context of the narrator's deprivation gives us to see once again a liminal experience of the subject, experiencing the margins, as much as the margins make the subject of his experience: a fairly representative picture of the Beat culture, showing in particular, in the words of Marcel Labine, 'a degraded social status' (2002: 61). But finally, Dean reappears and they escape into the night of San Francisco to listen to jazz. Here again, Sal feels at the end of the continent, at the end of something. His feeling is poetically reflected in the jazz music he listens to with passion: 'I never saw such crazy musicians. Everybody in Frisco blew. It was *the end of the continent* [emphasis added]; they didn't give a damn. Dean and I goofed around San Francisco in this manner until I got my next GI check and got ready to go back home' (Kerouac 2011: 161).

In this musical evening, the degradation of the narrator as a character seems once again to touch the depths of America, in which Kerouac finds a paradoxical illumination. This is another facet of the writer's Beat experience, in which, according to Labine, the declassed is pushed to 'look for a rhythm, an inner drive (the *beat* of the music), which will lead him to a state of *beatitude* close to enlightenment or the ecstasy of the musician' (2002: 61).[12] The end of the continent is no longer only geographical or social here; it is also spiritual. The end is therefore no longer strictly speaking an end, but becomes at this moment of the story a realization, an illumination.

Shortly after the musical evening in San Francisco, Dean's wife, Camille, demands that Sal leaves. The end of the continent is therefore for him once again as much personal as geographical. It is, moreover, an extra-diegetic end, since Sal Paradise's second voyage seems to lose all its meaning in relation to the narrative promise that has engaged the reader since the beginning of the novel.

## The Third Trip, from the Human Condition to the Human Comedy

The following spring, Sal Paradise undertakes another crossing of America. This is his third major journey and it corresponds to Part 3 of the novel. At this stage, the 'end of' motif conveys both the tragic and absurd nature of the characters' life experiences, if not of their whole existence. This is illustrated when things go badly with Camille and Dean is thrown out. Now it seems that it is Dean who has come to the end. He evokes the 'end of himself' to his friend by telling him how he injured his thumb during an argument with his mistress Marylou:

> 'That was months ago – after you left. She finally married a used-car dealer, dumb bastit has promised to kill me if he finds me, if necessary I shall have to defend myself and kill him and I'll go to San Quentin, 'cause, Sal, one more rap of any kind and I go to San Quentin for life – that's *the end of me* [emphasis added]. Bad hand and all.' He showed me his hand. I hadn't noticed in the excitement that he had suffered a terrible accident to his hand. (Kerouac 2011: 168)

The registers that Kerouac implements always oscillate between the spiritual and the absurd. Dean has lost his heroic stature. He is more of a tragicomic buffoon. However, he still fascinates Sal, who nicknames him 'the holy goof' (ibid.: 176). Later, they decide to enjoy the night of San Francisco. Once again, it is while listening to a frenetic jazz concert that the narrator feels he is at the end of the continent:

> Galatea and Marie were there, and it took us five minutes to realize it. Whoo, Frisco nights, *the end of the continent* and *the end of doubt* [emphases added], all dull doubt and tomfoolery, good-by. Lampshade was roaring around with his trays of beer; everything he did was in rhythm; he yelled at the waitress with the beat; 'Hey now, baby baby, make a way, make a way, it's Lampshade comin your way,' and he hurled by her with the beers in the air and roared through the swinging doors into the kitchen and danced with the cooks and came sweating back. (Ibid.: 183)

Here, the end of the continent is explicitly linked to Sal's personal state (and, we can infer, Kerouac himself in the real-life experiences that informed the novel). The end of the continent is mysteriously the end of a doubt, a doubt between the comic and the absurd. It is with the meaning – or nonsense – of life that the narrator is confronted. His doubts are amplified by the collapse of Dean's leadership stature. In the absurd, by dint of burlesque repetitions, and accounts of the misadventures of the characters that only lead to dead ends, comedy is born. The human condition becomes a human comedy. A little later, Dean and Sal decide to go back east. They arrive in Denver, where, during a drunken night, Dean steals a car; the following morning they get up early to move this stolen vehicle away from their lodgings:

> [Dean] stumbled out of bed, wearing just his jockey shorts, and we got in the car together, while the kids giggled from the windows, and went bouncing and flying straight over the hard alfalfa-rows at *the end of the road* [emphasis added] whomp-ti-whomp till finally the car couldn't take any more and stopped dead under an old cottonwood near the old mill. 'Can't go any farther,' said Dean simply and got out and started walking back over the cornfield, about half a mile, in his shorts in the moonlight. (Ibid.: 202)

The laughing children, the undoubtedly exaggerated car stunt and the onomatopoeic 'whomp-ti-whomp' expression evoke a comic strip world. The narrative again takes on a comical aspect, emphasized by the image of Dean 'in his shorts in the moonlight'. The 'end of' motif returns a little later when

Sal thinks he sees a police car arriving to arrest them, then realizes that it is only a taxi. The phrase 'end of his life', earlier used for Dean, is now used for himself:

> Early-morning traffic began, every car looking like a cruiser. Then we suddenly saw the cruiser coming and I knew it was *the end of my life* [emphasis added] as I had known it and that it was entering a new and horrible stage of jails and iron sorrows. But the cruiser was our taxi, and from that moment on we flew east. (Ibid.: 203)

The two friends manage to escape from Denver, headed for Detroit. This third part of the work recalls the picaresque novel, a point emphasized by Xiaoxiao Shi (2017) and supported by Jean-Yves Tadié's definition of the picaresque as 'open-ended, almost endless stories in which heroes from the working classes make a long journey; they cross many social environments, and get by thanks to their intelligence and cunning' (1982: 22).[13] Ann Charters also sees the picaresque genre in the adventures of Kerouac: 'Jack really had the experiences he described in his books, but when he turned them into picaresque adventure he gave them a sense of excitement by heightening his own emotional response to situations, at the same time as minimizing his own role as participant in the action' (1994: 292).

The idea of 'end of the road' returns on the Denver–Detroit route, through the fantasy that Sal attributes to Dean of finding a woman 'at the end of the road':

> The boys fell asleep and we talked and talked all night. It was remarkable how Dean could go mad and then suddenly continue with his soul – which I think is wrapped up in a fast car, a coast to reach, and a woman at *the end of the road* [emphasis added] – calmly and sanely as though nothing had happened. 'I get like that every time in Denver now – I can't make that town any more.' (Kerouac 2011: 209)

Sal is staged here as a witness to Dean's life, as he drives and talks tirelessly towards the destination that they have provisionally fixed. Sal sees Dean's life through the allegory of a race down the road, 'wrapped in a fast car', with only another border or 'coast to reach', and 'a woman at the end of the road'. Sal speaks about his friend's values in this passage. The pleasures of speed and fleeting attachments to women, which eventually amount to brief sexual experiences, seem to be all that hold Dean Moriarty's life. Implicitly, Sal invites us to see his friend, his hero as a being who begins to arouse pity. Finally, the real end of the road is to be found further east, in Detroit. Kerouac describes the marginalized people who have reached the end of the road, as, one might think, the end of their life, an irreversible path. Once again Kerouac offers a picture pitiful, sordid and paradoxically fantastic:

> Beat Negroes who'd come up from Alabama to work in car factories on a rumor; old white bums; young longhaired hipsters who'd reached *the end of the road* [emphasis added] and were drinking wine; whores, ordinary couples, and housewives with nothing to do, nowhere to go, nobody to believe in. If you sifted all Detroit in a wire basket the beater solid core of dregs couldn't be better gathered. The picture was Singing Cowboy Eddie Dean and his gallant white horse Bloop, that was number one; number two double-feature film was George Raft, Sidney Greenstreet, and Peter Lorre in a picture about Istanbul. (Ibid.: 222)

The narrator, without including himself completely, is involved in this crowd at the end of the road. Moreover, in this expression, the idea of an 'end' is no longer differentiated between the personal and the terminus of the road: 'The people who were in that all-night movie *were the end* [emphasis added]' (ibid.: 222).

Here again, Kerouac's obsession with the image of the end is revealed, with ending becoming an absolute experience. Kerouac not only experiences being beaten, but takes the metaphor further by describing himself and his friend as the end itself. They are 'ended', we may say. Through danger, a characteristic element of the travel narrative, Kerouac makes a detour through the epic register, a social epic, in a society that has normalized consumption where, ultimately, the liminal experience takes place through material deprivation.

## The Fourth Trip and the End, a Redemption at the End of the First Half of the Century

Part 4 of *On the Road* involves a new crossing of America for Sal. Dean is too busy with his love affairs and material troubles to follow him. But he eventually comes to meet Sal in Denver, and appears like 'a burning shuddering frightful Angel' (Kerouac 2011: 236). They take the road to Mexico together with their friend Stan. The idea of the end is again expressed by Dean:

> 'Ah,' sighed Dean, '*the end of Texas, the end of America* [emphasis added], we don't know no more.' It was tremendously hot: we were all sweating buckets. There was no night dew, not a breath of air, nothing except billions of moths smashing at bulbs everywhere and the low, rank smell of a hot river in the night nearby – the Rio Grande, that begins in cool Rocky Mountain dales and ends up fashioning world-

> valleys to mingle its heats with the Mississippi muds in the great Gulf. (Ibid.: 249)

Once again, the end is geographic; but while previous physical endings have been defined by the east or west coasts of the US, this ending is latitudinal: the southern border with Mexico. The narrator considers the end of the road before arriving in a city in Mexican territory; it represents the limit at the end of which the characters will find 'the magic land'. It is no longer a question of existential doubt, material annihilation, sordid experience of the margins, but rather of ultimate redemption. The end of the road is providential, it is represented by the geographical change of scenery and the dream of discovering new landscapes: deserts, tropical forests, mountains, plateaus.

> We gazed and gazed at our wonderful Mexican money that went so far, and played with it and looked around and smiled at everyone. Behind us lay the whole of America and everything Dean and I had previously known: about life, and life on the road. We had finally found the magic land at *the end of the road* [emphasis added] and we never dreamed of the extent of the magic. 'Think of these cats staying up all hours of the night,' whispered Dean. 'And think of this big continent ahead of us with those enormous Sierra Madre mountains we saw in the movies, and the jungles all the way down and a whole desert plateau as big as ours and reaching clear down to Guatemala and God knows where, whoo! What'll we do? What'll we do? Let's move!' (Ibid.: 251)

This idea of magic, associated with the end of the road, was also clearly present when Sal was in Los Angeles before leaving with Terry for the cotton fields: 'the fantastic end of America.' It is also found when Sal arrives in New York at the end of his first trip, 'the fantastic hoorair of New York' (ibid.: 96). But the magic is no longer a spark of happiness in a sweltering post-war America, it is no longer a bittersweet daydream, nor an irony of fate. The magic of Mexico, in Kerouac's prose, is new and generous. It is characterized by excessive natural elements: 'this big continent ahead of us with those enormous Sierra Madre mountains.' They feel that the sufferings are behind them. On the road to Mexico City, the characters decide to make a stopover, with the intention of spending the day using cannabis and hanging out with women involved in prostitution. The motif of the end of the road returns yet again, and it is a pathetic 'hard, hard road' that the narrator recalls, seeing in retrospect the road they have travelled, at the end of which they find a brothel: 'Still we couldn't sober up and didn't want to leave, and though we were all run out we still wanted to hang around with our lovely girls in this strange Arabian paradise we had finally found at the end of the hard, hard road' (ibid.: 265).

The pathetic aspect is reinforced by the slight irony of the purpose of their quest. Sal's character senses that something is wrong with the ending, which he exaggerates with his idea of an 'Arabian paradise', to which he adds the epithet 'strange'. Sal calls himself Sal 'Paradise' in a light-hearted, if not self-deprecating way, given the conditions of his journey. The idea of the end of the trip to Mexico returns a few pages later. The motif is raised here in a strictly geographical context, summarizing the distance travelled from Denver, in a Mexican setting with biblical touches:

> *The end of our journey* impended [emphasis added]. Great fields stretched on both sides of us; a noble wind blew across the occasional immense tree groves and over old missions turning salmon pink in the late sun. The clouds were close and huge and pink. 'Mexico City by dusk!' We'd made it, a total of nineteen hundred miles from the afternoon yards of Denver to these vast and Biblical areas of the world, and now we were about to reach *the end of the road* [emphasis added].
> 'Shall we change our insect T-shirts?'
> 'Naw, let's wear them into town, hell's bells.' And we drove into Mexico City. (Kerouac 2011: 274)

Once again, from Mexico, when the narrator evokes the end of the road, it is with a retrospective view. The narrator invites us to become aware of the distance travelled. Although the novel *On the Road* does not present a particular denouement, it is built in an evolution, during which the narrator finds himself at the end of the road. The retrospectivity of Sal Paradise's discourse, turned towards the past, prepares the end of his adventures – and of his novel – by reporting on the achievements made, in terms of kilometres, misadventures, liminal experiences. Finally, the motif of the end of the road is used here to designate the Mexican city that Kerouac and his friends will have finally found at journey's end.

> This was the great and final wild uninhibited Fellahin-childlike city that we knew we would find at *the end of the road* [emphasis added]. Dean walked through with his arms hanging zombie-like at his sides, his mouth open, his eyes gleaming, and conducted a ragged and holy tour that lasted till dawn in a field with a boy in a straw hat who laughed and chatted with us and wanted to play catch, for nothing ever ended. (Ibid.: 275)

At the end of the book, the phrase 'end of' is still used in a temporal context in a letter to Dean from his estranged wife and her children:

> I do want Sal and his friend to come and live on the same street ... I know you'll make it but I can't help worrying – now that we've decided everything ... Dear Dean, it's *the end of the first half of the century* [emphasis added]. Welcome with love and kisses to spend the other half with us. We all wait for you. [Signed] Camille, Amy, and Little Joanie. (Ibid.: 279)

It is no longer a question of a spatial end but precisely of a temporal end. The motif of the story, that of a century, concludes here the motif of the end often used in geographical terms. The second half of the twentieth century remains for the characters. The end here no longer invites us to look behind at the many kilometres already travelled. It is an opportunity to seize, to envisage the future.

## Conclusion

The motif of the end of the road/continent/America reflects an evolution of the level of meaning, register and genre. We first observe a geographical end, then a social one understood as a lowland of America, then a personal one, then a semantic end in itself, then an end confused with the characters who identify with the end. The register of nostalgia is more closely associated with the romantic ends introduced at the beginning of the novel. The spiritual and the absurd sometimes merge in the search for a Beat Revelation. This quest ultimately leads to the absurd – to what we call a human comedy, a picaresque ending. Finally, at the novel's conclusion, the more or less spiritual Mexican experience gives way to a nostalgic and romantic ending, marking the end of the first half of the twentieth century. The rhythm of *On the Road* can be read through the chronological analysis of this motif. The fresco of the human condition that the author aims for in his novel is largely based on the skilful use of this motif as a counterpoint to the image of the road. The end of the road can be understood as a shadow, an 'off the road' that contributes to the density of Kerouac's work. Kerouac carries a vision of the end of the road/continent/America that is at once realistic, symbolic and metaphorical, which reveals his whole literary approach as a writer. He offers a novel of the human condition through the eyes of a mid-twentieth-century traveller in the United States, by setting up a certain number of rites of passage whereby awareness of the human condition can be shaped. All of Sal Paradise's adventures take on meaning because at the end of the road he will tell the story. Rather than a traveller, Kerouac was essentially a writer-traveller, according to the accepted expression in French *écrivain-voyageur*. This expression refers to the category of writers who travel, but whose main profession remains that of writer. Kerouac found his place in this world by shaking up social norms and experiencing the margins of the America of his time, which were not his destination but the *sine qua non* of his work, like his plan, recounted at the beginning of *On the Road*, to become a professional writer. This makes Kerouac's novel a meaningful work for a generation in search of its bearings, as well as a significant work for study in the fields of literary theory and creative writing practice. Indeed, it sets out the important relationships of an author with his work project and his profession.

**Martin Wable** is a French PhD student specializing in travel literature and creative writing. His research focuses on representations of space, travel and identity in twentieth-century non-fiction narratives, with particular emphasis on post-war literature. In addition to his scholarly work, he is the author of several poetry collections exploring themes of geography and place. His collection *Géopoésie* (Cheyne, 2015) was awarded the Prix de poésie de la vocation. He has also published *Terre courte* (2020), a hybrid work blending poetry and prose, and *Les Étoiles* (2024), a collection of poems. His interdisciplinary approach combines literary analysis, cultural theory and geography, offering new insights into how post-war travel literature reflects broader social and ideological transformations throughout the twentieth century.

## Notes

1. The Penguin Modern Classics edition from 2000 has the same pagination.
2. There is a thirteenth occurrence in the 'original scroll' manuscript of *On the Road*, where 'end of the road' appears seven times in total (Kerouac 2007).
3. The following is a full list of noun forms associated with the phrase 'end of' appearing only once in the *On the Road*, Penguin Essentials edition (2011): *end of the* [metro] *line* (11), *end of the country* (13), *end of each little street* (15), *end of town* (29), *end of field* (87), *end of a* [vineyard] *row* (91), *end of something* (92), *end of the land* (96), *end of the second trip* (100), *end of the regular chain-gang run* (121–22), *end of me* (168), *end of his song* (181), *end of a doubt, all dull doubt and tomfoolery* (183), *end of all innumerable riotous angelic particulars that had been lurking in our souls all our lives* (189), *end of my life* (203), *end of it* (208), *end of the summer* (224), *end of every street* (232), *end of the stick* (237), *end of Texas* (249), *end of our journey* (273), *end of the first half of the century* (279).
4. On the subject of Kerouac and dispersion, or at least doubt, we might mention here the quotation from Albert Camus by Omar Swartz at the beginning of his book *The View from On the Road: The Rhetorical Vision of Jack Kerouac* (1999): 'It is good that from time to time we know doubt, for it provides us with the seriousness we need. We despise judges who never doubt and heroes who have never trembled. —Albert Camus, *Between Hell and Reason*.' This quotation shows that doubt is rhetorically fruitful for

an author who is the subject of the enunciation of his novel, since the reader recognizes themselves in the enunciator who confides their doubts.

5. One of Kerouac's main biographers, Ann Charters, writes of Neal Cassady, on whom the character of Dean Moriarty is based: 'Neal lived out the archetype of the image, but years later Carolyn [Neal's wife Camille in the book] was convinced Jack always carried a sense of guilt for creating his image of Cassady in *On The Road*, convinced he was partly responsible for Neal's decline' (Charters 1994: 333). This remark allows us, from the point of view of creative writing, moreover, to notice how an author of an autobiographical novel can be caught between reality and fiction, the fictive process having redefined their own life.
6. In this regard, we can cite Xiaoxiao Shi's thesis (2017), which compares Jack Kerouac's *On the Road* and Gao Xingjiang's *Soul Mountain*, both of which she defines as novels of the soul.
7. The most striking example of his tendency to mix dream and reality is undoubtedly his lesser-known 1959 book, *Doctor Sax*.
8. Lévi-Strauss's original French text is as follows: 'On conçoit généralement les voyages comme un déplacement dans l'espace. C'est peu. Un voyage s'inscrit simultanément dans l'espace, dans le temps, et dans la hiérarchie sociale.'
9. This is Kerouac's own expression.
10. The original French text is as follows: 'Il s'agit de réduire l'individu-confondu-avec-le-sujet au social. Parce qu'en effet le sujet est social. Et que l'individu est censé être l'anti-social, l'incompatible – au lieu que la collectivité n'existe que s'il existe. Mais un romantisme de la masse y est substitué. Stratégie du pouvoir Puisqu'on parle en son nom: "Le travail de l'œuvre d'art est social à travers l'individu, sans que celui-ci ait par là conscience de la société: peut-être d'autant plus qu'il en est moins conscient."'
11. Labine's original French text is as follows: 'ils cherchent le secret d'une Amérique autrefois sacrée, celle des pionniers et des pères fondateurs.'
12. Labine's original French text is as follows: 'chercher en lui un rythme, une pulsion intérieure (le beat de la musique), qui le mènera à un état de béatitude proche de illumination ou de l'extase du musicien.'
13. Tadié's original French text is as follows: 'on appelle picaresque des récits ouverts, et quasiment sans fin, dans lesquels des héros issus de classes populaires font un voyage de longue durée; ils traversent de nombreux milieux sociaux, et s'y débrouillent grâce à leur intelligence et à leur ruse.'

## References

Adorno, Theodor. 1974. *Théorie esthétique*. Paris: Klincksieck.
Charters, Ann. 1994. *Kerouac: A Biography*. New York: St. Martin's Press.
Kerouac, Jack. 2007. *On the Road: The Original Scroll*. New York: Penguin.
———. 2011 (1957). *On the Road*. New York: Penguin.
———. 2012 (1959). *Doctor Sax*. New York: Penguin.
Labine, Marcel. 2002. *Le Roman américain en question*. Montréal: Québec Amérique.
Lévi-Strauss, Claude. 2013 (1955). *Tristes Tropiques*. Paris: Plon.
Meschonnic, Henri. 2009. *Critique du rythme: Anthropologie historique du langage*. Lagrasse: Editions Verdier.

Shi, Xiaoxiao. 2017. 'Romans de la quête multiple: "La Montagne de l'âme" (Gao Xingjiang) et "Sur la route" (Jack Kerouac)', unpublished PhD thesis. Toulouse: Université Toulouse 2.

Swartz, Omar. 1999. *The View from On the Road: The Rhetorical Vision of Jack Kerouac*. Carbondale: Southern Illinois University Press.

Tadié, Jean-Yves. 1982. *Le roman d'aventures*. Paris: Presses Universitaires de France.

# CHAPTER 6

# 'A Strange Kind of Limbo'

## *Navigating the Unknown in Contemporary Anglophone Female Travel Writing*

Gemma Lake

### Introduction

As Rebecca Solnit puts it, 'It is the movement as well as the sights going by that seem to make things happen in the mind, and this is what makes walking ambiguous and endlessly fertile: it is both means and end, travel and destination' (Solnit 2014). If we define walking in the way that Solnit does then every step that we take is a voyage into the unknown: when we put one foot in front of the other, we enter a shifting terrain of unpredictable events that takes form in ways that remain fundamentally unknowable in their totality. Contrary to its traditional usage as the action of moving from place to place (*Oxford English Dictionary* n.d.), walking cannot be treated solely as a 'means' of 'travel' by which to reach an 'end' or 'destination', but as an 'end' or 'destination' in itself that remains forever open and ambiguous. Solnit does not, as a result, discover her destination at the end of her journey but remains at its centre, caught up in, and oriented by, its perpetual generation. Always there are 'ambiguous and endlessly fertile' ways to move through the world and be moved by it in return – to affect and be affected. This sense of tension in walking between affecting and being affected is captured by Brian Massumi in terms of freedom and constraint. For Massumi, bipedal activity conveys the 'freedom, or the ability to move forward and to transit through life' even though the reality of that freedom is constrained by 'gravity . . . balance, and a need for equilibrium' – the reason he describes walking as 'controlled falling' (2015: 12). This idea of walking as 'controlled falling' constitutes an apt metaphor for the way some contemporary female travel writers engage with the interval between moving and being moved, affecting and being affected. Let us consider *In Ethiopia with a Mule* (1968) by Dervla Murphy (1931–2022): attuning to the body in the bipedal mode of walking allows the Irish adventure travel writer to evoke, emphatically, a sense of freedom to move (or affect) that is constrained to some extent by a susceptibility to feelings of being moved. Irish American travel writer Rosemary Mahoney (1961–), another contemporary female travel writer, highlights a similar kind of tension between freedom and constraint in *Down the Nile: Alone in a Fisherman's Skiff* (2007) but through the rhythm of her rowing stroke.

Both convey the ambiguous and relational potential of affect through the dynamic image: a diagram of sorts for summoning up the body from a mobile character across different aspects of experience – thought, sensation, language. This site of ambiguity is endlessly fertile: the very ambiguity of walking or rowing feeds into an unpredictable potential for things to happen differently, so that to each dynamic image there corresponds the movement of affect (intensity). Filtering into their dynamic images is embodied knowledge or knowledge gained from and through the body, via feeling modalities – visual, auditory, tactile, olfactory, gustatory and kinaesthetic (or proprioceptive). The embodied basis of movement tinges their perceptual present in the liveliness of the writing process, amid shifting circumstances: a firelit *tukul* (or traditional Ethiopian home) alive with conversation in Murphy's case (1994: 38) or a boat on the 'shimmering and swirling' water of the Egyptian Nile in Mahoney's (2008: 230). As a result of their crossing of spatial and temporal boundaries, they enter into a liminal world of in-betweenness, an in-betweenness that becomes ever more complicated as the reader embarks on the journey with them. Victor Turner likens the existence of such a liminal entity to 'being in the womb, to invisibility, to darkness, to death, to bisexuality, to the wilderness, and to an eclipse of the sun or moon' (1996: 95). According to Turner, the liminal entity is 'neither here nor there' but 'betwixt and between the positions assigned and arrayed by law, custom, convention, and ceremonial' (ibid.). Their ambiguity enables them to 'elude or slip through the network of classifications that normally locate states and positions in cultural space' (ibid.). It seems to be precisely in their inhabiting liminal space by virtue of regularities of movement that Murphy and Mahoney engage with affect, or what Gregory Seigworth and Melissa Gregg describe as 'those intensities that pass body to body (human, nonhuman, part body, and otherwise) in those resonances that circulate about, between and sometimes stick to bodies' (2010: 1). Both are united in their attempt to make sense of their everyday lived experiences through their felt memories of thinking in action – immersed in eventful, rhythmic activity, in concrete landscapes and landscapes of the mind – and yet their dynamic images are shaped by very different experiences of affective intensity.

What this chapter explores, by means of affect theory, is a way of articulating the complexity, multiplicity and diversity of experiences of affective intensity in female travel writing through body motion. In the context of this collection, my chapter probes the role of body movement in female travel writing to argue three points: first, that the ambiguous nature of body movement in *In Ethiopia with a Mule* and *Down the Nile* echoes the ambiguity of an intensively felt dimension or affective atmosphere; second, that Murphy's and Mahoney's dynamic images illuminate the relational character of the affective encounter, blurring the boundaries between the self and the other, the perceiver and the perceived; and third, that the dynamic images proliferating across the texts cultivate an openness in bodies to other ways of thinking, knowing and doing gender. Whereas some approaches to female travel writing focus on how language reinforces gender norms, roles and identities – one quietly pervasive study being *Discourses of Difference* by Sara Mills (1991) – in this chapter, I draw on the flux and multiple becomings of identity in the genre through attention to contemporary theories of affect, especially the version underpinned by Deleuze and Guattari, elaborated upon by Brian Massumi, Kathleen Stewart and Sara Ahmed among others. Whereas a gendered discourse analysis of female travel writing, as demonstrated by Mills, focuses on specific communication styles relative to men's, an exploration of female-authored travel texts through the lens of affect encourages deep engagement with the diversity of everyday lived experience and the shifts in affective atmosphere that might arise from the spaces in between and problematize social categories. Applying to female travel writing these principles of affect, in this chapter, I focus on the ambiguous openness of bodies on the move.

The aim of this chapter is to reassess what Susan Bassnett calls the 'multi-focal' gaze in female travel writing (2002: 240) through attention to what Kathleen Stewart calls a 'bloom space' (2010: 340): a space of emergences, transitions and transformations. Stewart defines a 'bloom space' as that which 'can catch you up and then deflate, pop, leave you standing, a fish out of water. Or, same thing, it can catch you in its moves' (ibid.: 342). Whereas Bassnett's 'multi-focal' gaze focuses on the power of observation and the privileged position of the observer, Stewart's 'bloom space' drives the reader towards movement in a world that is unsettling, ambiguous and intensely interactive. As I see it, the female travel text is replete with troublesome bloom spaces where bodies orientate themselves through attunement to embodied movement and the movement of affect. Though ostensibly accounts of two very different environmental, social and cultural milieus, both texts, as I will argue in this chapter, draw attention to a flexible dynamic by way of movement that problematizes the imaginary boundaries between self and other. Like walking, the concept of affect is rather elusive and means different things to different people. For contemporary affect theorist Brian Massumi, affect is 'continuous like a background perception that accompanies every event, however quotidian', which tends to emerge as the 'perception of one's own vitality, one's sense of aliveness, of changeability' (2002: 36). In this sense, affect can only ever be felt as an atmosphere that is hard to grasp. Of course, as Highmore notes, this is never a smooth process, for the everyday is also rife with 'fissures in the stream of constancy' made up of 'interruptions and irruptions: a knock on the door, a stubbed toe, an argument, an unexpected present' (2011: 1). Murphy's and Mahoney's dynamic images of their bodies in motion are, in many ways, replete with such 'interruptions and irruptions' or affective moments, whether they are walking or rowing. In their accounts of their idiosyncratic journeys over land or water, Murphy and Mahoney invite us to engage with an affective intensity or atmosphere brought about by a particular way of moving through space.

Take for instance *In Ethiopia with a Mule*: the act of walking forces the author and touring cyclist out of her comfort zone and into the unknown territory of hiking, illuminating an intense relationship with the world. During the winter of 1966–67, Murphy spent three months walking from the Red Sea shore to Addis Ababa via Aksum, the Semien mountains, Gondar, the northern shore of Lake Tana and Lalibela. Looking back on her journey, Murphy admits that she was unprepared for the 'walking-tour' and unaccustomed to uphill hiking with a heavy backpack that weighed 'fifty pounds' (1994: 8). Perhaps this declaration is unsurprising: the author is best known for travelling on two wheels, and *Full Tilt: Ireland to India with a Bicycle* (1965) bears testament to her enthusiasm for this mode of transport. In an interview with Tim Hannigan, Murphy states that *Full Tilt* was years in the making, spurred on by a childhood ambition at the age of 10 and prefaced with numerous bicycle trips to Wales, England and continental Europe (Hannigan 2021: 132). Thus, Ethiopia, a land of high plateaus and rugged mountains that is ill-suited to travelling on two wheels, posed a unique set of challenges to the avid cyclist. The physical challenges of hiking seem, in many ways, to facilitate analysis of other movements underway in Ethiopia, including the Eritrean Liberation Front, an armed resistance movement that emerged in reaction to the annexation of Eritrea under Haile Selassie, the emperor of Ethiopia from 1930 to 1974 ('Eritrean People's Liberation Front' 2023). The '*shifta* (bandits)' associated with the Eritrean Liberation Front are a dominant presence in *In Ethiopia with a Mule* (1994: 10; original emphasis), affecting the way in which Murphy moves through and navigates the environment. In regions where *shifta* have been observed, she is more often than not escorted and forced to accept her 'loss of freedom' (ibid.: 77) to wander unaccompanied. In many ways, her journey was facilitated by the eldest granddaughter of the emperor, Leilt (Princess) Aida Desta, and British consuls in Asmara and Addis Ababa, respectively, who supported her at different points along the way. Her surrendering of certain freedoms in exchange for others presents a conundrum that occasionally leads to feelings of frustration. The legacy of the Italian invasion – an invasion to which Selassie led resistance while living in exile in the West (Milkias 2006) – also has an impact on Murphy's movements. In some of her interactions, she is mistaken for being Italian and male, and detects an attitude of suspicion born out of what she considers to be an 'Italian-bred fear of *faranjs*' or foreigners (Murphy 1994: 36). At times, her heightened sensitivity to local attitudes sets in motion a range of different feelings, from angst to disdain to vulnerability. As a result, some of her most cherished moments are spent travelling in her own time, at her own pace – even though such moments are not entirely free of constraint.

Years after Murphy took to the mountains of Ethiopia on foot, in 1998, Mahoney set off down the Egyptian Nile in a fisherman's skiff, rowing some 120 miles from Aswan to Qena. This was a journey that Mahoney had been considering for 'more than two years' since her first trip to Egypt, and she arrived with years of rowing experience (Mahoney 2008: 6). In spite of her well-laid plans, however, Mahoney devotes a good portion of the book to her time in the city of Aswan, where purchasing a boat becomes a major stumbling block to finding freedom on the water. Mahoney experienced Egypt in the wake of the Luxor massacre, an Islamist terrorist attack carried out on tourists during the presidency of Hosni Mubarak ('Hosni Mubarak' 2025). Although the attack does not define her journey, her awareness of the increased police presence in Luxor and Aswan does have a bearing on her mobility in some of her experiences. In these popular tourist destinations, Mahoney develops a heightened sense of self-consciousness as a Western woman travelling alone in an 'Islamic nation' (Mahoney 2008: 4) where, from the point of view of the narrator, the local women act according to a very different set of social standards. In areas beyond Luxor and Aswan where tourists 'must . . . in theory at least, be accompanied by a police convoy', Mahoney avoids attracting 'unwanted attention from the Egyptian police' by confining herself to the middle of the river under the guise of a local fisherman (ibid.: 15). It is in this liminal state that the paradox of movement emerges: Mahoney cannot move without being moved by the forces working with and against her body, oars and boat.

Thus, Murphy and Mahoney navigate the cultural complexities of their respective environments through movement in very different ways and on very different levels, providing ample opportunities to explore the uneven terrain of affect – a terrain upon which processes of perception play out in various ways, according to the lived scale of the everyday. Together, their divergent accounts allow us to engage with female travel writing through movement and mechanisms of affect in infinite ways – according to how the words move us in some way and depending on what we as readers bring to the process of meaning-making. What body movement makes accessible to us, and hence knowable, is the feeling of recurrent suspense with every step and swing forward. Massumi (1995: 85) emphasizes the importance of the body in the registration of affect, asserting that 'Intensity is embodied in purely autonomic reactions most directly manifested in the skin – at the surface of the body, at its interface with things'. In other words, we embody intensity, involuntarily, without conscious effort. For Massumi, embodied intensity arises in linguistic expression through 'resonation or interference, amplification or dampening' (ibid.: 86). What makes the dynamic image resonate with or amplify intensity in moments of suspension in the reading process remains an open question. In order to convey the sense of movement, Murphy and Mahoney have recourse to undeveloped affective forces that have the potential to resonate, unpredictably, with the reader.

## Rhizomatic Connections

The dynamic image is in this sense a meandering concept for thinking about the body that is not governed by any fixed norm or image of itself, but a manifestation of the creative tension between what is knowable and what is unknowable. The dynamic image constitutes a kind of knowledge of the world that appeals to the polyvalence of everyday lived experience, encounters with other bodies, an awareness of which is reached through attunement to the affective atmosphere underlying the eventfulness of movement expressed in the written word. The eventfulness of movement is what Massumi might describe as '[a] whole field of potential composed of lines of potential action, and their enabling or disenabling conditions' (2015: 194). Indeed, the potential of the dynamic image to affect and envelop the reader in sensations of movement depends on the capacity of bodies to move and be moved. As such, the dynamic image of the body itself in motion is open to the sway of lived experience, vulnerable to the practices of appropriation and interpretation beyond the printed page. In essence, bodies on the move are at once connected and autonomous, governed by an unpredictable affective unfolding. The ambiguous nature of body movement carried forward by the female traveller to expression lingers in dormancy, awaiting reactivation (or germination) by the reader. This latent ambiguity in language allows the reader to conceive of the world in movement wherein the potential lies: the excess of affect or intensity is a by-product of the capture of experience that presses for creative expression. Depending on how we attune to the mystifying nature of affect, the ambiguity that arises in the writings of Murphy and Mahoney at the level of the body in motion can lead us into different spaces, across different temporalities.

Even if we do not have a share in the movement underway or identify with the subjective viewpoint of the author, we as readers become caught in a web of relations by means of modulation, interference, resonance. This recalls the concept of the rhizome developed by Deleuze and Guattari (2013) to relate a network of connections that is non-linear. Rhizomes are underground plant stems with shoots and roots that enable them to spread laterally. Deleuze and Guattari compare rhizomatic thinking with the vertical, hierarchical structure of arborescent thinking. Whereas arborescent thinking follows a well-trodden path, rhizomatic thought pulls us in different, sometimes unexpected, directions and provides a means of making connections. Viewed through the concept of the rhizome, the dynamic image in female travel writing plays out its relational potential by encouraging lateral, creative thinking in acts of reading. Like the rhizome, the dynamic image can grow horizontally, sometimes generating affective bloom spaces whose qualities are vaguely felt as an affective atmosphere. Feeling for an atmosphere through the image of the body itself in motion depends on how the processual event of movement appeals to our senses or stimulates imagination. To adapt Solnit's phrase, this is what makes the dynamic image so 'ambiguous and endlessly fertile': it is an expression of variations in affective experience that moves and resonates with the event to come.

The dynamic image can provoke or stimulate feelings in complex and sometimes contradictory ways, one such example being the Debarak scene in *In Ethiopia with a Mule* in which Murphy summons up memories of walking along a track, north of Gondar, which 'rose and fell over a succession of yellow-green or grey-brown ridges that were like the immobilised waves of some mighty ocean' (1994: 129–30). The simile here captures a sense of suspended movement in the landscape through the repetition of colour, shape and form that seems to arise from the rhythm of steps and the rhythm of thinking. Reflecting on the undulating terrain through attunement to the felt memory of walking, accompanied by the rhythmic four-beat gait of her pack-mule Jock ambling alongside, Murphy recalls entering into a peaceful trance state that carries a suspense of its own: 'After walking for a few hours through this sort of terrain one's whole being seems soothingly involved in the gentle rhythm of regular climbs and descents' (ibid.: 130). As the wavelike regularity of the mountain ridges pulls Murphy along with the rhythm in the affective landscape of Ethiopia, there is a sense that the narrator has become harmoniously wedded to the world. In Stewart's words, she has become finely attuned to 'a world of affinities and impacts that take place in the moves of intensity across things that seem solid and dead' (2007: 127). Her attunement to a world of 'affinities and impacts' or relational affects in the flow of 'intensity' between her body and the seemingly inanimate terrain leads us to regard the process whereby Murphy makes sense of the world through movement. It is through movement perception to which vision contributes that the rolling expanse of ridges comes to represent a space filled with life as well as a site of freedom that is constrained to some extent by geologic refrains and the 'gentle rhythm of regular climbs and descents'. Recalling Massumi's idea of walking as 'controlled falling', Murphy moves through the landscape by 'playing with the constraints' (2015: 12). Ultimately, then, movement is the key site of Murphy's exploration of the interplay of freedom and constraint that is brought to expression through a process of rhythmic attunement to a transitional in-between state or passage through space, hinged on the feeling of anticipation of a next step.

The trancelike interlude, brought on by the soothing rhythm of putting one foot in front of the other, fosters a certain open-ended in-betweenness that translates movement into a kind of contemplative reverie or meditation, a meditation interrupted by an unexpected encounter with human nature. 'On the crest of one ridge I saw my first highland funeral – two men carrying a *shamma*-wrapped corpse on a simple bier, followed at a little distance by women keening professionally' (Murphy 1994: 130). As Murphy crosses the threshold of awareness in social space, other movements emerge in the exchange of affect, leading to a feeling of conspicuousness: 'Tears were streaming down their cheeks, but when they saw Jock and me their weeping and wailing stopped and for ten minutes they stood excitedly speculating, while the corpse went on its way unmourned' (ibid.: 130). The sense of fluidity evoked by the words 'tears' and 'streaming' is filled with motion. Comparably, the word 'speculating', which is a borrowing from the Latin *speculātiōn-em*, meaning reflection or conjecture (Collins Latin Dictionary 1997), is characterized by a reflective posturing that is awkward and stilted but not altogether motionless. The encounter between Murphy and the 'speculating' crowd yields to the fluidity of speculation that the corpse went by 'unmourned', speculation premised on assumptions about the cultural politics of emotion and (in)authenticity of grief. Murphy responds to the encounter with a different sense of loss or kind of grief in relation to the space of 'unpeopled remoteness' through which she claims to have traversed for the last 'fortnight' (1994: 130). The dramatic shift from 'unpeopled remoteness' to a track that seemed 'crowded' and consumed by 'the "civilisation" of mass-production' (ibid.) prompts her to reflect mournfully on the encounter with the funeralgoers as an interruption to her sense of inconspicuousness in her peaceful passage through the landscape.

In this affective bloom space where unsuspected possibilities for new ways of thinking, being and acting might arise, Murphy is brought into a different embodied and connected relationship with the world, a world in which, according to the narrator, it was 'considered frightfully non-U' (or inappropriate) for a foreigner to 'walk instead of riding' and the 'absolute bottom' for a foreigner 'personally to lead a pack-animal' (1994: 47). That and local concerns about the 'impossibility of anyone walking through the Semien Mountains, least of all a solitary woman' (ibid.: 60) made walking alone with a pack mule an act of nonconformity in an area where Murphy moved to a different beat.

The tension between freedom and social constraint arising from, to borrow a phrase, 'the field of collective differential attunement' (Massumi 2015: 117), makes itself felt as part of an event, an interruption, imbued with an affective tonality or atmosphere that feeds into and shapes how Murphy responds to the situation. This moment of revelation and reflection requires Murphy to embody contradictory qualities simultaneously, thus producing an unresolved conflict between the feeling of freedom to move through a seemingly 'unpeopled' landscape, uninterrupted, and the feeling of constraint in the space of sociality.

Unresolved tension between freedom and constraint can affect what happens next, as the following example illustrates. In her diary entries on the stretch between Nefasit and Asmara, Murphy focuses on how becoming active makes her feel in a way that influences how she interacts with the world: 'The process of converting a cyclist into a hiker is being rather painful. ... At the moment my shoulder muscles are fiery with pain and – despite the most comfortable of boots – three massive blisters are throbbing on each foot' (1994: 14). Indeed, being 'out of training' (ibid.: 14) creates a very particular set of bodily challenges for Murphy that constrains what her body can do. Concretely, the challenge of becoming a hiker is painfully learned by Murphy as discomfort and struggle. The concreteness or felt quality of walking has its own anticipatory rhythm under 'fiery' shoulder muscles and 'throbbing' blisters, which seems to affect what happens and how the atmosphere is angled. Murphy, recalling the four-mile stretch to Ghinda, writes: 'Everyone stared at me suspiciously and only once was my greeting returned ... One doesn't resent such aloofness, since surprise is probably the main cause, but I soon stopped being so unrewardingly amiable' (ibid.: 14). Indeed, the mood she arrives with on foot would seem to affect what impression she receives of others insofar as bodily pain enacts the intensity that gives shape to the raw material of her experience, as well as direction to her social discomfort. By implication, not only is walking physically painful for Murphy but it is also a form of social expression that enables her to extract value from the affective encounter. The impression of movement functions as an intense force that accumulates over time, based on the capacity to affect and be affected, or what Deleuze terms *affectus* (1990: 220), which predisposes Murphy to act and react in particular ways. It is through the tension between freedom and constraint inherent in walking under the gaze of another that Murphy conveys an impression of becoming conspicuous, a conspicuousness that stops her 'being so unrewardingly amiable' and creates an atmosphere of uncertainty. In Murphy's experience of traversing this intense space of uncertainty, certain transmutations of feeling seem to be projected onto certain bodies through movement, as well as proximity, according to the value invested in anticipation.

Massumi refers to sensation as 'the feeling of anticipation' or the 'registering of potentials' (2002: 92). In the case of Murphy, the sensation of walking corresponds to a certain anticipatory tone that seems to infuse the moment with a mood that seeps into the space the narrator comes to inhabit. Thinking through proximity of the other also allows Murphy to reflect on the process whereby walking under suspicion is constituted as the cause of her vulnerability: 'Already I notice a difference between cycling and walking in an unknown country; on foot one is even more sensitive to the local attitude and one feels a little less secure' (1994: 14). Murphy is moved by the memory of the affective encounter, and that memory triggers a feeling of vulnerability in that it requires her to become 'more sensitive' to the world as she opens herself up to others. That vulnerability is intensified by an affective force that makes movement matter to her in very specific ways, in her contact with a world where travelling on foot heightens her sensitivity to perceived criticism or judgement and stimulates feelings of insecurity. The act of walking brings the otherwise fleeting and unattainable cyclist into a field of relational intensity where she is immersed in an affective bloom space, a space that, according to Stewart, 'demands collective attunement and a more adequate description of how things make sense, fall apart, become something else' (2010: 340). The other comes into her subjective expression, retrospectively, as perceived 'aloofness', aligning the act of walking in an 'unknown country' with an awareness in hindsight of angst, disdain, vulnerability. To orientate her body towards others Murphy has to become 'sensitive' to the effectively immobilizing factors, such as feelings of vulnerability, that affect what her body can do.

The memory of walking compels a movement outwards towards objects and the bodies of others, thus imparting what Guattari refers to as a 'subjective pluralism' or 'multiplicity within oneself' (1996: 216). This pluralism provides multiple trajectories of transitory affects that can transform into different degrees of mobility within a complex relational dynamic, but it is also rooted in the body's sense of movement, action and location. The uncertainty that Murphy imparts based on her experience of the relational encounter gives way to an atmosphere of uneasiness. For Stewart, 'Everything depends on the feel of an atmosphere and the angle of arrival . . . on the dense entanglement of affect, attention, the senses, and matter' (2010: 340). Everything we feel as readers depends, therefore, on multiple factors, thus making the atmosphere conveyed through movement infinitely irreducible. The infinitely irreducible atmosphere conveyed by Murphy through a web of 'affect, attention, the senses, and matter' seems to arise from the feeling of transition from one experience to another, while under the gaze of another. Affective atmospheres, in Ben Anderson's terms, are 'a class of experience that occurs *before* and *alongside* the formation of subjectivity, *across* human and non-human materialities and *in-between* subject/object distinctions' (2009: 78; original emphasis). Drawing our attention to propositions of space and time – '*before and alongside*', '*across*' and '*in-between*' – Anderson helps us to think of an atmosphere as a process that is underway in the background, as well as a mode of becoming attuned to the internal and external forces at play in a world of alternative potentials. Attending to the atmosphere in Murphy's writing becomes, therefore, an exercise in acceptance: we must learn to accept the excess of affective intensity in the uncontainable movement and change of everyday life.

## Inhabiting the In-Between

Whereas Murphy uses the dynamic image of walking through the Semien (or Simien) Mountains in the Ethiopian Highlands to re-engage with everyday affective life, Mahoney focuses on the rhythmic task of rowing (see Chapter 2 of this collection for an earlier, male-authored account of this rhythmic mode of travel). The Egyptian Nile is for Mahoney a space in flux, a bloom space, where mobility translates into feelings of freedom and constraint. The 'deeply soothing' quality of rowing that imparts a sense of 'freedom and isolation' (2008: 130) for Mahoney is juxtaposed with 'an element of surrender in the exercise, an active acknowledgement of how breathtakingly tiny and helpless [she] is in the greater scheme of things' (ibid.: 9). The felt memory of rowing holds Mahoney in a state of indeterminacy between 'freedom' and 'helpless[ness]' intensified by the threat of uncertainty that accompanies every swing of the oar in an unpredictable environment. Mahoney navigates this terrain of uncertainty by making use of manifold aspects of her environment such as the 'simple costume' that looked 'vaguely like what an Egyptian fisherman would wear' and 'vernacular boat' (ibid.: 218). These props enable her to inhabit and move through otherwise inaccessible spaces, and yet the freedom of movement arises from the constraint of the other: inhabiting a stretch of the Nile where she had 'neither seen nor heard of any foreigners . . . unaccompanied by an Egyptian captain or of a single woman, Egyptian or otherwise, operating a boat on the river' (ibid.: 13) requires her to erase the attuned differences of the other. On the surface, Mahoney draws her reader into an apparently straightforward encounter, an encounter whereby the mobility and presence of the traveller involves the immobility and absence of other bodies. Of presence and absence, Mahoney writes: 'No one cared; no one suspected. I was so inconspicuous that I had begun to feel almost invisible . . . I felt like a spy – outwardly one thing, inwardly another. I was a misfit in the best possible way' (ibid.: 238). Part of the assumption here is that the sense of in-betweenness or liminality engaged by Mahoney involves an awareness in hindsight of the indifference of the other: it is assumed that one can inhabit different social skins. And yet, the experience of inhabiting another skin makes Mahoney very aware of her bodily capacity to affect. According to Mahoney, no one shares her preference for rowing in Egypt where her 'pastime' is 'tantamount to laying railroad tracks or digging graves or tarring highways' (ibid.: 74). Mahoney's identification of rowing as a privileged leisure pursuit transforms this image of freedom into an image of difference. This transformation is explicit in Mahoney's central question: 'Why would a foreign woman do the work that rowing required when she could hire any Egyptian man to do it for her for a pittance?' (ibid.). Mahoney recognizes here the relationships of power at play in Egypt: the question becomes not so much what rowing is but what it does.

The contrast set up between different degrees of freedom and constraint on the Egyptian Nile by way of body movement draws our attention to how certain affective situations either enable or constrain the individuals involved. This tension can also be detected in the 'Hoda and the Women' chapter of *Down the Nile* in which Mahoney recalls a conversation with felucca owner Amr Khaled: 'In Egypt, a Western woman would never truly be a woman, nor did she quite approach the status of a man' (2008: 114). Whereas Egyptian women 'had to be isolated, contained, and controlled' by their male counterparts, 'a Western woman ... was more like ... a pleasant but irrelevant animal like, say, a peahen or a manatee. It was like moving through a strange kind of limbo' (ibid.: 113–14). In this 'strange kind of limbo', Mahoney assumes the role of the liminal entity whose transformative nature enters the scene in different ways in connection with questions of gender, sexuality and culture, challenging our normalized, possibly privileged positions. 'Not Nubian, not Muslim, not Egyptian – these facts disqualify me entirely from the female category', writes Mahoney (ibid.: 113). Like the liminal entity characterized by Turner, Mahoney inhabits the space betwixt and between the gender ideologies that apparently locate 'Nubian', 'Muslim' and 'Egyptian' women. Her heightened awareness of her liminality disrupts the binary oppositions that regulate the structural and social codes of her reality: 'It struck me that [Amr] was wearing a dress and I was wearing trousers and a man's hat' (ibid.: 94). Mahoney, based on feedback from the event, creates the conditions for redefining normative dress codes within Western constructions of gender. A traditional dress-like garment worn by both men and women in Egypt, the *jalabiya*, becomes couched in affects that Mahoney carries forward.[1] As a result, the idea of a man wearing a 'dress' perceived as a female garment and a woman wearing 'trousers and a man's hat' engages and challenges Mahoney, and by association her readers, to reconsider the gender codes of the Western cultural tradition in a dynamic and fluid process. Visual markers such as the *jalabiya*, trousers and hat that link Amr and Mahoney to attributes normatively considered male or female in Western cultures encourage a move beyond the inadequate binary between active subjects and passive objects. The focus shifts towards bodily mobility and the capacities of bodies in relation to the other bodies they come into contact with, indicating a complexity and fluidity in affective mediated relations.

Variations on the theme of fluidity recur throughout *Down the Nile*, an example being a scene in the 'Not Floating but Flying' chapter in which Mahoney recalls surrendering the protection of Amr for the freedom of movement (and thought). In this passage, rowing becomes a state of being that inspires a 'feeling of physical lightness' or 'weightlessness', like 'drifting on air' (2008: 218). In the same way as walking through the Ethiopian Highlands transports Murphy into a meditative state, the memory of rowing, uninterrupted, permits Mahoney to suspend herself in a world in which the material and the immaterial become increasingly fluid. The idea that the body can exist in an environment as a kind of free-flowing capsule seems to unleash what seems to be a mixture of elation and guilt: 'I rowed this way, in an overexcited trance, for approximately an hour before I began to feel guilty about straying out of Amr's sight' (ibid.: 122). Mahoney exists temporarily in this 'overexcited trance' state until she is constrained by a 'guilty' feeling that leads her to hang back in 'a deeply shaded spot that seemed safely secluded' until Amr makes a reappearance (ibid.). The feeling of guilt weighs heavily upon Mahoney, affecting her ability to move freely through the landscape. As the journey progresses, this affective bloom space where realities grow like rhizomes in unpredictable ways becomes all the more complex: 'Most fishing boats stayed within range of their villages, rowing back and forth across the river, or up and down the local banks, like horses in a corral. . . . Few boats went off on journeys like mine' (ibid.: 221). This simile of frustrated horses in a corral exposes the social injustices connected to and enabling Mahoney's passage through a space filled with resistances and compromises. Thus it becomes clear through movement that her reality is just one among many. Like the lateral-spreading rhizome root, her dynamic image is emergent across a complex network of relations, presenting the reader with an unfolding problem – a problem that lies in reality.

Deleuze, drawing on Henri Bergson, views reality in terms of the 'virtual' and the 'actual' (1994: 239). The virtual is associated with potentiality, emergence and creativity, the actual with the concrete conditions of experience (a distinction further explored in relation to travel in Chapter 9 of this collection). These two aspects of reality are not mutually exclusive but constantly influence and shape one another in the process of perception. In regard to *Down the Nile*, the virtual is deeply embedded in the actual by way of body movement, and this feedback loop, in turn, informs our actual experience of reading. Drawing on the triangle concept developed by Gaetano Kanizsa, Massumi describes the capturing of the virtual aspect of lived experience as a 'kind of emergence', an 'emergence effect' that, like the triangle, has 'no actual lines connecting the corners' and 'yet it appears' (2015: 185). The triangle acquires 'actual' shape in the same way as the dynamic image unfolds in the mind of the reader, but only by means of a virtual realm that remains in a constant state of flux. The virtual then can only ever be partially captured or actualized as an impression of aspects of lived experience, depending on how the body (or triangle) appears to us. Like Kanizsa's triangle, the dynamic image emerges in different ways, inviting fluid and mobile interaction.

## Conclusion

This chapter has argued that the dynamic images of *In Ethiopia with a Mule* and *Down the Nile* emphasize the affective intensity of everyday life in bloom spaces where bodies exist in suspension. Murphy's and Mahoney's diverse renderings of affect attune writer and reader to the intensively felt dimensions of movement through relations of freedom and constraint, putting into question the boundaries between the self and the other, between the perceiver and the perceived. Seeking out intensities, tensions and varieties of liminality, I have argued that Murphy's and Mahoney's dynamic images of their bodies on the move press against the binary differences reified into formulations of patriarchy by drawing attention to other ways of thinking, knowing and doing gender. As this chapter has shown, affects or intensities, however fleeting, give way to ambiguities in their writings, which, at times, makes it difficult to extrapolate both to the category of female travel writing, particularly when taking into account their liminal status within their respective environments and relational fields. Through the lens of affect, I have explored how female travel writing might be examined beyond the various limitations of gendered discourse analysis, as adopted by Sara Mills, which, in my view, discounts gender's multiple becomings within various configurations of patriarchies that challenge the very dualistic thinking that underpins and sustains them. The at times disorientating loop of the affective realm unravels in *In Ethiopia with a Mule* and *Down the Nile* through rhythmic, often unpredictable, resonances of the body itself in motion, expressed in language qualified by emotion, putting into question structures of power that rely on linear structures of thought.

Bloom spaces in both unsettle social norms through connection to a rhizomatic network in which the reader is implicated. As figures moving through and navigating the 'unknown' in a 'strange kind of limbo', Murphy and Mahoney expose the paradox of movement as well as identity, both of which are contingent on different aspects of reality – virtual and actual. Offering glimpses of the affective cosmos of their realities, Murphy and Mahoney express their felt memories of walking or rowing through dynamic images framed by the physical, spatiotemporal dimensions of their unique environments, which are not entirely knowable. Through somatic attunement to the liveliness of their heterogeneous encounters with other bodies, both nurture a sensitive awareness of how affective situations either enable or constrain the individuals involved. In doing so, they point to the disruptive openness of affect while also cultivating ambiguous openness to the unknown. We, as readers, are encouraged to add to the ongoing modulation of affect in this dynamic process of attunement, which, like the process of walking or rowing, is in a constant state of flux and transformation. There is undeniable tension throughout both texts, which encourages us to rethink the concrete conditions of experience as well as the notion of movement. The contemporary female travel writer's gaze becomes less 'multi-focal', as Bassnett puts it (2002: 240), but multi-tonal: shifting through dynamic feedback loops that create a certain affective atmosphere or the illusion of a three-dimensional effect, a Kanizsa triangle, oriented towards alternative paths through differential attunement to their individual yet deeply relational situations. Based on my own unique reading of their written commentaries of movement, there is hope in the constellation of affective flickerings in their dynamic images – a galvanization of deeper political complexities bound up with gender – for seeking out infinite mutually imbricated and rhizomatic opportunities to challenge, grow and flourish within the ever-shifting affective bloom spaces that emerge in their writings.

## Acknowledgements

Many people have contributed to my thinking on various issues concerned with gender identity in Anglophone female travel writing: Gilles Deleuze, Félix Guattari, Brian Massumi, Kathleen Stewart and Sara Ahmed among them. I would like to acknowledge Christine Berberich, the angel on my shoulder, who encouraged me to submit this chapter in the first place and who has remained a constant support throughout my studies, both as a peer and as a friend.

**Gemma Lake** is a PhD student of English literature at the University of Portsmouth, UK, writing on the experiences of contemporary female travel writers and the affective landscapes they traverse. Her approach has roots in contemporary affect theory and epistemological scholarship with particular focus on how affective forces shape social values. Her work is influenced by, among others, the writings of Gilles Deleuze, Félix Guattari, Brian Massumi, Kathleen Stewart and Sara Ahmed. Around her research, Gemma has worked as a freelance travel writer and editorial assistant for Rough Guides and a structural editor for Summersdale Publishers.

## Note

1. According to Mona Abdou in 'Jalabiya: Egypt's Iconic Kaftan' (2022), the *jalabiya* is a loose-fitting traditional garment worn by men and women in Egypt that remains true to its original design with only minor variance in cut, colour and accessories attributable to gender, region and economic status. For instance, it is common for an Egyptian fisherman exposed to the elements to wear a loose-fitting white *jalabiya* in the summer months to better regulate body temperature.

## References

'Affect'. 2022. *Oxford English Dictionary Online*. Retrieved 14 February 2023 from https://www.oed.com/view/Entry/3321?rskey=mXz6jY&result=1#eid.

Abdou, Mona. 2022. 'Jalabiya: Egypt's Iconic Kaftan', *Egyptian Streets*, 28 February. Retrieved 5 July 2024 from https://egyptianstreets.com/2022/02/28/jalabiya-egypts-iconic-kaftan/.

Anderson, Ben. 2009. 'Affective Atmospheres', *Emotion, Space and Society* 2(2): 77–81.

Bassnett, Susan. 2002. 'Travel Writing and Gender', in Peter Hulme and Tim Youngs (eds), *The Cambridge Companion to Travel Writing*. Cambridge: Cambridge University Press, pp. 225–41.

Deleuze, Gilles. 1990 (1968). *Expressionism in Philosophy: Spinoza*, trans. M. Joughin. New York: Zone.

———. 1994 (1968). *Difference and Repetition*, trans. P. Patton. New York: Columbia University Press.

Deleuze, Gilles, and Félix Guattari. 2013 (1980). *A Thousand Plateaus: Capitalism and Schizophrenia*, trans. B. Massumi. London: Bloomsbury.

'Eritrean People's Liberation Front'. 2023. *Encyclopedia Britannica*. Retrieved 20 March 2025 from https://www.britannica.com/topic/Eritrean-Peoples-Liberation-Front.

Seigworth, Gregory J, and Melissa Gregg. 2010. 'An Inventory of Shimmers', in Melissa Gregg and Gregory J. Seigworth (eds), *The Affect Theory Reader*. Durham, NC: Duke University Press, pp. 1–25.

Guattari, Felix. 1996. *The Guattari Reader*, ed. Gary Genosko. London: Blackwell.

Hannigan, Tim. 2021. *The Travel Writing Tribe*. London: Hurst.

Highmore, Ben. 2011. *Ordinary Lives: Studies in the Everyday*. London: Routledge.

'Hosni Mubarak'. 2025. *Encyclopedia Britannica*. Retrieved 20 March 2025 from https://www.britannica.com/biography/Hosni-Mubarak.

Mahoney, Rosemary. 2008 (2007). *Down the Nile: Alone in a Fisherman's Skiff*. New York: Back Bay Books.

Massumi, Brian. 1995. 'The Autonomy of Affect', *Cultural Critique* 31: 83–109. https://doi.org/10.2307/1354446.

———. 2002. *Parables for the Virtual: Movement, Affect, Sensation*. Durham, NC: Duke University Press.

———. 2015. *Politics of Affect*. Cambridge: Polity.

Mills, Sara. 1991. *Discourses of Difference: An Analysis of Women's Travel Writing and Colonialism*. London: Routledge.

Milkias, Paulos. 2006. *Haile Selassie, Western Education and Political Revolution in Ethiopia*. Youngstown, NY: Cambria Press.

Murphy, Dervla. 1994 (1968). *In Ethiopia with a Mule*. London: Flamingo.

Solnit, Rebecca. 2014 (2001). *Wanderlust*. London: Granta.

'Speculation'. 1997. *Collins Latin Dictionary*. Retrieved 24 March 2025 from https://search.credoreference.com/articles/Qm9va0FydGljbGU6MTY0NzI1Mg==?aid=279718</div.

Stewart, Kathleen. 2007. *Ordinary Affects*. Durham, NC: Duke University Press.

———. 2010. 'Afterward: Worlding Refrains', in Melissa Gregg and Gregory J. Seigworth (eds), *The Affect Theory Reader*. Durham, NC: Duke University Press, pp. 339–53.

Turner, Victor W. 1995 (1969). *The Ritual Process: Structure and Anti-Structure*. Piscataway, NJ: Transaction.

'Walk'. n.d. *Oxford English Dictionary*. Retrieved 12 March 2025 from: https://doi.org/10.1093/OED/8299973573.

CHAPTER 7

# Writing from the Ice

## *An Examination of Contemporary Travel Writing in Antarctica*

Kelly E. Hall

### Introduction

As a bumper sticker for sale in the McMurdo Station store clearly reminds those who work there, *Antarctica: It's a Harsh Continent*. The sticker features a lone explorer, his or her features completely obscured by a bulky parka and blowing snow. Guidebooks and articles are quick to point out that it is the highest, driest, windiest, coldest continent and inhospitable to human life. But for those who have been, Antarctica, or 'the Ice' as we call it, is a place of great adventure. I spent four austral summers at McMurdo Station, first as a vehicle operator, driving scientists and other workers out to airfields on the sea ice, and later as a cargo coordinator for a field camp. My time in Antarctica was filled with so many once-in-a-lifetime experiences and I can attest that visiting the frozen continent warrants the time, effort or expense it takes to get there. The desire to describe and share one's adventures while on the continent is a common reaction. Antarctica affords amazing opportunities that allow writers, both professional and amateur, to describe these experiences with the full understanding that what they are doing is interesting, enviable and rare.

The first Antarctic travel writing, composed during the early 1900s when European expeditions were mapping the continent and racing to the Pole, was limited to three basic forms: personal journal or letter, published account, and newspapers written on the Ice for the enjoyment of fellow explorers. Almost all writing was done on the move, or shortly after being on the move, as men were sailing to the continent, sledging across it or exploring it on foot.

Early personal writing and published accounts deal almost exclusively with the hardships of the experience and the men's reactions to it. As Sir Ernest Shackleton writes while man-hauling supplies across an ice floe in *South!*:

> By 2 a.m. we were on the march again. We wished one another a merry Christmas, and our thoughts went back to those at home. We wondered, too, that day, as we sat down to our 'lunch' of stale, thin bannock and a mug of thin cocoa, what they were having at home. All hands were very cheerful . . . One man wrote in his diary: 'It's a hard, rough, jolly life, this marching and camping; no washing of self or dishes, no undressing, no changing of clothes . . . sleeping almost on the bare snow and working as hard as the human physique is capable of doing on a minimum of food.' (Shackleton 2000: 120–21)

Even when facing long days pulling sleds and eating reduced rations, Shackleton and his men maintain their good spirits. Contemporary writers continue this tradition, and prefer to describe the good, the funny, the amazing or the sublime. They don't often complain. They realize how privileged they are, understanding that the rewards of going to such a remote and rare place outweigh the negatives. A breaching humpback will always be stunning, no matter how rough the seas.

Unlike travel writing about other continents, Ice writing is arguably free of the taint of colonialism or colonial guilt. Antarctica has no indigenous population and almost no permanent residents.[1] Everyone who goes there is a visitor. No longer exclusive to relatively young, male explorers, Antarctic writing is now produced by tourists, scientists and workers of all ages and gender identities.

This chapter examines these modern Antarctic visitors and their writing: the people who go there, the conditions under which they arrive, how they spend their time, how they compose their writing and the subjects they consider the most meaningful. Their purpose in writing is not merely to record or inform. They go to the Ice to experience something rare or profound and they write about it to have that experience acknowledged and validated.

## Those Who Go

There are two types of people who go to Antarctica: tourists, who generally see the edges of the continent from a cruise ship, and workers, who conduct scientific research or complete employment contracts as support staff at the various bases and field camps. The two groups do not often interact, and workers will report how nervous tourists can be when arriving for a brief station tour. In his book *Big Dead Place*, waste engineer Nicholas Johnson recounts, in his typical humorous style, his encounter with a cruise-ship tourist:

> One time I hopped out of my machine as I was making my rounds and asked one of the tourists a question about the ship. She laughed ner-

> vously and ran to catch up with the group as if I had asked her for spare change. I had approached her on my best behaviour, but as I recovered from the sting of rejection, I realized I must look a mess, lurking in filthy clothes amid a terrific array of clanging metal and hydraulic voodoo at this lawless outpost on the lip of the world.
>
> Where tourists are slain and their fat burned as fuel to heat our bowling alley. (Johnson 2005: 64)

Workers often have jaded opinions of tourists, saying that they visit Antarctica only to say they've been, that they are rich or spoiled, that they did not do it the 'hard' way or that they didn't truly experience the continent.

However, the two groups do share a love of adventure, the desire to explore a vast and barren continent and a longing for rare experiences. As worker Michelle Endo rightly claims, 'Regardless of what you're on the continent to do, once you've made it, you're part of the exclusive albeit sometimes weird community of misfits' (2022). This very exclusivity is a great part of the appeal.

If tourists and workers have similar reasons for going, the two experiences are vastly different. Most tourists arrive on package cruises. A few will sail from New Zealand or Australia to view the Ross Ice Shelf, but the majority will depart from Argentina or Chile to view the Antarctic Peninsula. These trips last between ten and thirty days, usually taking four days to cross the notoriously rough seas of the Drake Passage. Travellers spend their time viewing icebergs and wildlife, which can include penguins, seals and whales. Trips can include science or history lectures, sea kayaking excursions, visits ashore to smaller research stations or penguin colonies and occasionally the chance to camp overnight. The voyage takes little preparation beyond the normal amount for any trip, such as buying the proper cold-weather gear, booking flights and obtaining travel insurance. Meals are prepared by chefs on board the vessels, participants sleep in ensuite staterooms and housekeeping is done by the ship's crew.

Tourism has been increasing exponentially over the past few decades, barring the recent decline due to COVID-19. The IAATO (International Association of Antarctic Tour Operators), which was formed by independent companies, has been recording tourist numbers since their formation in 1991. They reported 6,700 tourists in the 1992–93 season and (before the start of the pandemic) had projected 78,500 for 2019–20, which was the highest number ever, and an increase of over 20,000 people compared to one year earlier (McClanahan 2020). By contrast, only 10,000 workers go to Antarctica each year.

Those who take a cruise come from various countries and backgrounds and are either independently wealthy or have spent time saving to go on what they describe as the 'trip of a lifetime'. A trip can cost from USD 6,000–50,000, depending on the length of the cruise, the type of vessel and the amenities provided. Because of the high cost, tourists usually only visit Antarctica once. When asked why they want to go to Antarctica, many tourists will respond that they want to see pristine scenery and wildlife, but the most popular reason is that tourists want to say they've been to all seven continents (McCulley 2018). As first-time tourists, they are most likely to write about what they saw while cruising to or around the continent, or what life is like aboard ship.

By contrast, those who go to Antarctica for research or employment must undergo more extensive preparation and be willing to stay for longer periods of time. US-based scientists travel to Antarctica to perform research in biology, zoology, meteorology, astronomy, oceanography and geology. They must secure permission for their projects years in advance with the National Science Foundation, seek additional funding when necessary, obtain leave or sabbatical from their institutions, recruit vital support personnel and organize their logistics. They will have the job-related task of writing on the move while conducting their research: lab reports, animal counts, project updates or operational instructions. They may also choose to write personal reflections, all while moving around the continent to their field camps or research locations.

The employees who support this research outnumber the scientists by a factor of eight to one. These are the cooks and dining attendants, janitors, vehicle operators, mechanics, logistics coordinators, postal workers, cargo specialists, housing staff, supply handlers, pilots, flight crews, medical personnel and communication specialists who keep the stations running. Obtaining a job in Antarctica is difficult and can take many years. Most positions are filled by those who worked previous seasons, or those who have the benefit of a referral from a friend or colleague.

Workers will go there for two weeks to a year, with the majority going during the austral summer season from October to February. Preparation is a months-long, intense process involving doctor and dentist visits, a long list of medical exams and blood tests, online training and reams of paperwork. Before heading to Antarctica, they will spend half a day trying on thermal underwear, fleece outerwear, wind pants, waterproof boots and parkas to find the best fit for their job needs.

Those heading to stations on the Peninsula will sail through the Drake Passage like most tourists, and the workers heading to the South Pole or stations on the opposite side will fly from New Zealand on military C-17 or LC-130 aircraft. There is only one cafeteria at each station and most food arrives on the continent canned, boxed or frozen. Workers sleep in dormitories or temporary field-camp shelters with shared bathrooms, and everyone helps with housekeeping duties. The mandated work week is fifty-four hours.

Stations and camps are small and tightly knit. McMurdo Station, one of the three permanent US bases, can have one thousand people during the summer season, but fewer than two hundred during the winter. Temporary field camps will often have fewer than ten. Predominantly white and leaning to the political left, US stations average 25 per cent women, and are inclusive to the LGBTQ community and senior citizens.[2]

Education levels will vary. There is a running joke about how there are more workers with PhDs in the galley than in the science lab, since doctors, lawyers and professors have sometimes taken dishwashing jobs for the experience. Keri Nelson, who has spent over fifteen years on the Ice at the three US stations, explains in a video for school children that she worked in news reporting and as a registered nurse before going to McMurdo to work as a janitor (2022).

Antarctica is a love-it-or-hate-it place. Employees will do a 'one and done' season, or come back multiple times, often moving into higher and better-paying positions. Scientists will seek additional grants to continue their research in later years. Of course, not everyone who works in Antarctica writes about their experience, but those who do, especially in professionally published formats, are often those who have completed multiple seasons. Furthermore, transferring to other jobs often allows them to travel to new places on the continent that they can then discuss in their writings. When asked why they go back, workers will sometimes joke, 'Well, it's not the food.' They quote the familiar, self-deprecating mantra 'You go the first time for the adventure. The second time for the money.[3] And the third time because you have nowhere else to go.'

The actual reasons are more complicated. Some employees prefer a nomadic life, and will spend as much time on the move while *off* the continent as they do on it. They enjoy having free time between seasons for personal travel or the flexibility of working at national parks when they're not on the Ice. Many will say they go back for the people. Antarctica is a place where lifelong friendships form, where romantic relationships begin (or end) and where most people thrive in an unusual place. Conditions may be challenging, but those who work there are great problem solvers who enjoy working with others to find solutions. As Endo states, 'there is a huge culture of respect across occupations and mutual understanding that everyone has something to bring to the table' (2022). Lastly, returning to Antarctica allows opportunities to see more of the continent. Someone who comes for the first time as a janitor to McMurdo may return for the chance to work further afield at the South Pole or a field camp.

## The Conditions of Writing

People who go to Antarctica are constantly on the move. No matter where the journey begins, or how one arrives, travel to the continent takes time and usually involves more than one type of transport. Tourists spend their time moving from place to place on larger vessels, zodiacs and kayaks, and often set foot ashore for further exploration. Workers move around station or off station for their jobs, heading to the airfield or more remote field locations. All recreation trips and hiking excursions take place away from base. Professional writers, many of them on writing grants or assignments, will often visit more than one location. Sara Wheeler, author of *Terra Incognita* (1998), and the first woman to receive a US Antarctic Writer's Grant, details visiting six of the major research bases and several smaller camps in the seven months she spent there.

As with most travel writing, writing from Antarctica typically takes place after the fact. Tourists are not able to describe kayaking beneath a glacier until they are back on their ships, and vehicle operators are not able to describe driving a terra bus for the first time until they are back on station. People in Antarctica are likely to describe the extraordinary, as discussed further below, and they want to write about these experiences as quickly as possible. This is particularly true of social media posts, where photographs are taken in the moment but the commentary is composed later.

Writing outside is nearly impossible. The low temperatures and high winds make writing by hand difficult, and laptop and tablet batteries run down very quickly in the cold. Finding the time and space to write can be difficult. Tourists report having busy days, even when their ships are in transit on open water. Their blogs outline all the activities on board, such as history and wildlife presentations, photography classes, game nights, karaoke and yoga. Megan Jerrard, a frequent traveller and blogger, informs other travellers that they likely won't have much time in the ship's library:

> Before we boarded the ship I was quite excited about the prospect of the library. A quiet cozy place for reading and relaxation, it has an extensive selection of polar books and DVDs, magazines, and reference materials, and we often stopped in for a quick scan of world newspapers printed daily. But while I had grand plans of reading in my down time [*sic*] in reality, there was too much else to do. (2017)

Employees and scientists at camps and stations are equally busy. They work ten- or twelve-hour shifts and many work overtime. Travellers can bring supplies if they choose to handwrite letters, postcards or journals, and many bring laptops and tablets, though internet access is limited. Computer labs and Wi-Fi are available on cruise ships, though they often incur an additional cost. Larger scientific bases will have small computer labs and ethernet cables in the dorms, but the internet speed is a frequent complaint from those who work there. McMurdo doesn't have a consistent amount of bandwidth for Facebook and other forms of social media, and the South Pole station only has internet for a certain number of hours each day.

If finding the time and means to write is difficult, so is finding the space. Many people go to Antarctica hoping to 'get away from it all', but the irony is that travellers are always around people on the Ice. Whether on a cruise or at a research facility, everyone eats in the same cafeteria, shares the same social spaces and sees the same people day after day. Virginia Woolf famously discussed the need for a room of one's own, but here only a very few people have single rooms. Tourists will often share a stateroom with partners or family, but some report sharing with strangers in multi-bunk rooms to save money. In her travel blog *Ottsworld*, Sherry Ott reports the difficulty of finding time to be alone:

> The beauty of the Ross Sea trip was that it is a small ship with under 50 passengers. However, you are with those same people for an entire month for every waking moment. Every meal is a social experience. You are expected to go eat in the dining room and eat with everyone. This means there is a social pressure to chat with people during that time. I know that doesn't sound like a big deal – but some mornings I woke up and just desperately wanted to eat without having to chit-chat. . . . I'm a very social person, but there were few moments of alone time on a ship of that size. (2022)

She also discusses her assigned roommate, with whom she gets along very well, but states that 'you don't always get that lucky at the roommate roulette' (ibid.).

This holds true at research bases as well. Workers are assigned roommates if they don't arrive with a friend or partner, and some unlucky first-year employees will have three roommates. Stations have day and night shifts, and workers will come and go from their rooms at all hours. There are lounges in the dorms, though these are often used for movie nights and other social events. McMurdo Station has a lovely, small library, but it is staffed by volunteers and only open a few hours each week.

Thus, personal writing is not an easy task. Crowded rooming conditions, lack of comfortable outdoor spaces and busy schedules force writers to create their own opportunities. When I was a vehicle operator, I would sometimes write in my journal while sitting in a vehicle alone, waiting for my passengers to land at the airfield. Other workers will find one of those quieter spaces and use headphones and a laptop. Anyone who wants to write must make the extra effort to do so.

## Rare Experiences

Travel writing composed while on the move in Antarctica covers almost all forms of general writing: poetry, prose, fiction and non-fiction. A reader can easily access articles, scholarly research, books, blogs, cartoons or social media posts. Tourists who go to Antarctica, especially those who write travel blogs and articles for a living, will frame their posts as personal experience, an instructional guide or a combination of the two. Often sponsored by travel providers, these posts are overwhelmingly positive and reverent in tone. Those who are recounting their adventures will write about how they got to Antarctica and how they spent their time, and may include photos of what they were able to see.

The travel blog *Live Hard, Travel Often*, founded by a Seattle-based married couple who travel professionally, currently has five informative articles posted about their cruise, covering the topics of how to visit Antarctica, the cost of the trip, the crossing of the Drake Passage, kayaking while on their cruise and what they call the 'ULTIMATE Polar Plunge' (Kylie and Scott 2022). Sherry Ott's *Ottsworld* blog gives extensive advice and includes ninety-two categories about her preparation and experience on two different cruises (Ott 2022).

Tone will vary slightly, but tourist writers share a common enthusiasm. Their opportunities are similar, but in experiencing them for the first time, they write with a sense of reverence or excitement. Nellie Huang describes sailing through the picturesque Lemaire Channel, also known as 'Kodak Gap', with a sense of wonderment:

> The bright sunshine bounced upon the glassy water surface reflecting off all the white surrounding me – the white iceberg crabeaters that floated past us, the white from the mammoth mountains that enveloped us and the white from the glaciers that slide off ice cliffs. It was a gorgeous morning, with the sun lighting up the entire landscape in a golden glow. (2012)

Dave and Deb, the founders of the travel website *The Planet D*, a top-ten influencer according to *Forbes*, describe their zodiac boat tour in Pleneau Bay with equal enthusiasm: 'Taking photographs of crabeater, leopard, and Weddell seals sleeping on ice flows while working our way through the maze of icebergs was one of the most thrilling experiences we have ever had' (Dave and Deb 2021).

Tourist blogs share similarities because their writers usually have the same goals. They want readers to enjoy and share their articles. They need 'likes' and comments to ensure continued sponsorship and they profit financially when readers choose a product or tour they promote.

By contrast, very little writing from scientists and employees is meant to be instructive. Workers and researchers are not expecting their friends or family to follow in their footsteps; thus, the majority of their personal writing is meant to inform others about where they are or what they experience. For example, Dr John E. Carlstrom, a professor of astronomy and astrophysics at the University of Chicago, describes his current project on cosmic microwave background:

> The 'Stage-4' ground-based cosmic microwave background (CMB) experiment, CMB-S4, consisting of dedicated telescopes equipped with highly sensitive superconducting cameras operating at the South Pole . . . will provide a dramatic leap forward in our understanding of the fundamental nature of space and time and the evolution of the Universe. CMB-S4 will be designed to cross critical thresholds in testing inflation, determining the number and masses of the neutrinos, constraining possible new light relic particles, providing precise constraints on the nature of dark energy, and testing general relativity on large scales. (Carlstrom 2002)

His homepage is meant to be informative for a general audience. He's not giving instructions at this point on how to conduct this type of research, and his findings will likely be published in an academic forum after his research is complete.

Not all writing is formal, published or posted online. Informal and spontaneous writing occurs on stations when there are performances or 'open mic' nights. The McMurdo Station Women's Soiree, held annually from 2007 to 2011, was organized by women on station as a talent show to raise funds for a local charity in Christchurch, New Zealand.[4] The Soiree featured female per-

**Figure 7.1.** Matt Davidson, 'A Brief History of Suffering in Antarctica', from *Freeze Frames*, 2008 (Morrisville, NC: Lulu Press), 52. © Matt Davidson.

formers, and, along with the requisite skits and dance numbers, also included poetry readings and original songs. It was a fun event and a valid fundraiser and was also a creative outlet for writers and artists who wanted the opportunity to perform. In 2009 the station held a writing contest called 'Romancing the Ice'. The two winners were Jana Dietrich, who wrote an essay about her personal experience and romantic relationship during a winter season, and Danny Gregory, who wrote a fictional story about a female worker at McMurdo who receives messages from a secret admirer. The prize for them and their two guests was a sled ride to a field camp and a gourmet dinner.

This less formal writing is a source of creativity and often an outlet for humour. Matt Davidson, who worked several years as an air-ground equipment operator, composed a cartoon each week during his summer seasons. His cartoons often poke fun at the station bureaucracy or those who work there:

The joke here is that clearly life in Antarctica is not so difficult now as it was for early explorers. McMurdo Station has hot showers, heated dormitories and a soft-serve ice cream machine in the galley. No one dies from hypothermia or scurvy.

Another cartoon by Davidson addresses current events, such as the week he was travelling back and forth to the air strip on the sea ice, and vehicles were

**Figure 7.2.** Matt Davidson, 'Beat the Heat', McMurdo Station, Antarctica, 2014. © Matt Davidson.

constantly getting stuck in the melting slush. While it was 42°F at McMurdo, cities in the northern US were having record low temperatures:

This beloved cartoon was shared by many workers with friends and family back in the states, who couldn't believe that an Antarctic summer was warmer than a northern winter. Loved ones in Minnesota could not get their cars started, but people at McMurdo were jogging in shorts.

Workers going to Antarctica for the first time will usually describe their accommodations, food, free time and especially their jobs. Scientist Michael Nayak, a principal investigator at the South Pole for LANDIT (Long-Duration Antarctic Night and Daytime Imaging Telescope) spends an entire chapter of his blog describing his work with the telescope and future plans for the project (2018). South Pole general assistant Kiell describes her work at the vehicle maintenance facility, which consists of long days spent moving fuel drums, chipping ice from the frozen floor and shovelling snow (2022).

First-timers are also likely to write about parts of the application or arrival process that will seem mundane in later years. Nicholas Johnson's *Big Dead Place*, which many workers consider a primer for working in Antarctica, details much of the red tape and bureaucracy inherent in the US Antarctic Program. With his typical dry humour, Johnson describes the initial briefing that all scientists and workers attend, even if they have been there previously:

> Soon after I arrived at McMurdo Station by plane that first summer, the station manager gathered the employees in the galley for an orientation . . . I only had one question: How long can I stand outside before I die? The station manager, instead, told us that if our neighbours were noisy, we should report them to the Firehouse. (Johnson 2005: 2)

Those who work at US stations generally have a good sense of humour about inefficient program management. Many first-time workers report similar astonishment that this orientation does not include directions on how to survive a skua attack, but instead includes instructions for the complicated process of filing an expense report.

Workers are most likely to write about experiences that are out of the ordinary, and often removed from their jobs. The *Antarctic Sun* recently featured an article that described a January performance of the 'McMurdo Nutcracker', a reimagining of the classic ballet that involved several departments (Lipuma 2022). Social media posts from McMurdo increase sharply after station-wide parties that feature music and costumes, like Halloween and the New Year's Eve concert called Ice Stock. South Pole workers will write about joining the '300 Club' in which workers sit in a sauna cranked to 200°F (93°C) and then run outside in the snow when the temperature reaches –100°F (–73°C). Jennifer Scotia posts about something she had not experienced in her twelve seasons at McMurdo: 'We had a partial eclipse this weekend and it was a magical event where we got to go out on the ice and look through welder's glass, homemade pinhole contraptions (in colour and upside down – so fun!), and big telescopes with filters. Such a fun community event' (2021).

The holiest of grails for an Antarctic worker is the boondoggle, an opportunity to leave the station to see something extraordinary. Usually parcelled out by chance or lottery, these trips are highly desired, very rare and a source of much jealousy. Examples include a day with the BBC diving crew, a helicopter ride to a field camp, a snowmobile trip to the Erebus glacier tongue, a trip to an ice cave or a visit to the Cape Evans historical hut. One of the most prized trips is the 'Sleigh Ride', a flight that allows a worker from McMurdo to visit the South Pole station. This involves a three-hour flight to the Pole, and then a quick thirty-minute visit to the station and the ceremonial barber-shop 'pole' that represents the bottom of the world. As Scotia describes her sleigh ride in 2011, it was the 'Best Boondoggle Ever!' Accompanied by photos, her blog marvels at South Pole station resources: 'It was startling and marvelous [*sic*] to see plants at the greenhouse! ... Wow, they have a great gym!' (2011). At this point her words and photos represent what many see or do when they are rewarded with this morale trip, but when her plane is diverted on the return trip, she gets to experience something completely unexpected. She lands at the CTAM (Central Transantarctic Mountains) field camp and ends up spending the night. Scotia writes about the warmth of her hosts and the impact of several new, unexpected visitors: 'We were fed tea and homemade ginger cookies ... Later on in this adventure I spent hours in the kitchen doing pots and pans to help out, since we doubled the size of the camp' (ibid.). She is driven on a snowmobile sled up the mountain for even greater views and gets to view the enigmatic, wind-shaped stones: 'The Rocks had an Easter Island feel to them, as if watching the eons of time pass without regard to the human race' (ibid.).

As diverted planes are a very rare occurrence, Scotia's blog shows how one rare boondoggle turned into two. In addition to visiting the Pole, she went to an even less-visited field camp and was able to experience what few others do. Anyone who has worked at McMurdo can easily understand why she would describe this trip as 'the highlight of my season' (2011). Anyone hearing or reading this story from Scotia would have been happy for her, but also ridiculously jealous.

## Reader Responses

As established, most Antarctic writing is positive in tone. People may complain verbally about the food or station management, but they don't want to leave a lasting impression of negativity. No matter how they arrived, those who go to Antarctica realize what a unique place it is and how lucky they are to be there. The majority of responses are equally positive, no matter the forum or the writer's goals. A recent Facebook post from McMurdo employee Steven Windsor provides a typical example. He describes his snowmobile boondoggle as 'the most exciting trip I've ever been on' (2021). Forty-six of the forty-nine replies express the respondent's individual enthusiasm for his adventure, with comments such as 'Live that adventurous life!' and 'Looks absolutely amazing!'

Reader response, most easily studied in social media posts, varies according to whether the respondent has been to Antarctica. Those who have not been see it as a grand adventure, something to which they aspire or something they enjoy vicariously. People who have been there will comment on their relationship to the post, having done the same thing, or remember something similar.

Tourist blogs are intended for other like-minded adventure-travellers. Informative, and even reflective, travel blogs do not expect reader response from those who have worked there. Respondents are most often those who wish to go, or those who have already been there as a tourist. Sherry Ott's informative travel blog *Ottsworld* has thirty responses to her post about how to prepare for an Antarctic cruise (Ott 2022). Twenty-six of these are from people who want to do something similar. One responds, 'Amazing photos and story! I love penguins so much, and this is very high on my list of things to do in my life.' Three posts are from those intending to go, such as April, who responds, 'You have answered so many questions! I normally don't get excited about a trip until I arrive, but your blog posts have gotten me overwhelmingly excited! I can't stop smiling because I know I am one step closer to the most pristine place in the world (and my 7th continent)! Thanks again!'

Four of the responses are from readers who have already been, such as Sandra Powers, who writes, 'I thank you for bringing back a rush of memories. The Ushuia [*sic*] rolling houses, the Beagle Channel, the confluence of the Atlantic, Pacific and Southern Ocean, the huge waves crashing over the entire ship, the nausea, and finally the incredible beauty, light, whites and blues of the Antartic [*sic*].' Powers's response is one of the longer ones, and she elaborates about her experience sailing to Antarctica in 1992 with a cast of colourful characters. Powers's only reference to Ott's post is her opening phrase, 'Loved your article', and the remainder of her long response clearly prioritizes her own experience. Powers specifically notes that she went on a cruise when they were not as popular or prevalent, and that she did it on a repurposed Russian research vessel. In writing a response to Ott's post, Powers magnifies her own experience, and subtly implies that her adventure was all the more rewarding because of the difficult means of travel.

Just as tourist blogs do not expect a worker response, the reverse is also true. Typical responses to workers' social media posts are from friends who are amazed by the scenery or the experience, but a significant number of responses come from fellow Ice workers. When the question was asked on the Facebook group *Ice Friends Helping Ice Friends*, 'What percentage of your Facebook friends are Ice friends?' the average response was 30 per cent.[5] Jeff King, who states that he hasn't worked on the Ice in twelve years, answers 35 per cent, and James S. Roth, who states that he spent most of the last decade at McMurdo, answers 81 per cent. Not only do these numbers speak to the friendships formed in a close community, but they also indicate the likelihood of an Ice friend responding to a post about Antarctica.

When Dan Schieffelin (2021) reposts a ten-year-old photo of a McMurdo volleyball game with everyone is dressed in crazy costumes, Melanie McDaniels remembers the event and states, 'Oh man that was so fun. Team shuttles was rockin'!' When Lori Haugen (2022) posts about her 'Santa ride', a boondoggle spent visiting field camps via helicopter to spread Christmas cheer, Rhoda Davison Bonneau responds, 'Lenny and I got to go on the Santa trip one year. Definitely a highlight of our Antarctic time.'

Those who go to Antarctica are members of a small community who share experiences not found elsewhere. Thus, responses to social media posts can be about reminiscence, as seen in Schieffelin's and Haugen's posts, or they can provide an opportunity for the respondent to highlight their own, possibly better, experience. A Facebook post from Michelle Tweed provides the perfect example. She includes a photo of sledges and a canvas Scott tent and describes her snowmobile boondoggle and hike through the IMAX Crevasse as 'One of the best days of my life. I still miss this place' (2017). Like Powers's response to Ott's blog, Ice workers will also respond to a post with more information about their own experience. Alan Cox replies that he had camped in that type of tent while there: 'Lived and worked out of those tents as flight mech on a R4D-7 (circa 1962-1963) Topo East Topo West USGS survey project!' Nick Majerus responds, 'I never made it there. I usually passed on the trips to the guys in the crew. I and other Foreman would go up into the Dry Vallies [*sic*] with "Woody" and check out the Camps there to see what needed to be repaired for the next season.' Majerus mentions both his magnanimity, in passing boondoggles on to others, and the fact that he had one of the rarest experiences in Antarctica. Only a very few select visitors get to view the Dry Valleys, and a trip there is rarer than a visit to the South Pole.

## Hubris and Humility

Just as people groom their social media personas, so too do Antarctic travellers want to shape how they are perceived. When asking fellow tourists why they were on her cruise, blogger Kate McCulley notes, 'Everyone had a bit of ego in their response. People were travelling here because they wanted to feel more important, more experienced, more special. They didn't just want to visit somewhere remarkable, they wanted to *become* more remarkable as a result of having visited' (2018).

Tourists who write about their rare experiences want to have those experiences acknowledged as something rare. Travel bloggers Alesha Bradford and Jarryd Salem have the requisite posts on their site about sea kayaking and how to pack, but they also detail their trip to the less frequently visited Vernadsky Station. They describe the small Ukrainian base as an important scientific facility where the hole in the ozone was first discovered, but also as a fun, quirky spot where ladies could trade a bra at the southernmost bar for three shots of the station's homemade vodka: 'It is a rare privilege to visit an active research station, and having the opportunity to explore one in a place as remote as Antarctica made the experience all the more rewarding' (2022). In composing this post, Bradford and Salem know they have experienced something rare on a continent that is already far from the beaten track.

Writers are equally keen to highlight when they have done something the hard way. While traversing the Drake Passage, tourist Neill Drake advocates for a rough crossing, linking his experience to what early explorers faced:

> This was the age of discovery. The age of exploration. They did it without Gore-Tex, without GPS, without satellite phones or up-to-the-minute weather reports. They did it in wooden ships without stabilizers or diesel engines. It's insane to think about what they accomplished in those times with how little they had by today's standards. Experiencing at least one rough Drake Passage crossing will put into perspective what these early explorers endured to make these discoveries . . . There is a sense of adventure and exploration that comes with '*surviving*' the crossing. While pretty much every single passenger on board will pray to Poseidon for a smooth sailing, I feel you should have at least one rough crossing . . . Antarctica is the most remote place on the planet. You should earn it. (2021)

Workers, like tourists, are not immune to the joys of bragging rights. Johnson correctly states that 'swaggering one-upmanship is common at every level of The Program' (2005: 92). Those who have been for more than one season, those who have spent a winter on the Ice or those who have been to the South Pole or Palmer Station will be able to describe, or respond to, an experience that first-time summer employees will not have. Those who are lucky enough to see remote camps, such as Jennifer Scotia in the example above, will be able to describe something many workers in Antarctica never see. Those like Sara Wheeler and Gabrielle Walker who go to Antarctica on NSF Artist and Writer

Grants can visit not just the US bases and field campus, but also other international stations.[6]

Going to Antarctica is a rare opportunity and seeing a calving glacier from a zodiac or peering into the Mount Erebus caldera from a helicopter is an even rarer one. Writing about a place that seems extraordinary to most readers, and then highlighting an even more extraordinary event within that trip, allows writers to magnify themselves and their experiences. They become an even more elite member of an already exclusive club.

Scientist Michael Nayak clearly knows how lucky he was when he describes his experience:

> In 2018, I was proud to be part of the United States Antarctic Program at [the] South Pole. Being here is a Venn diagram of steadily smaller and more unique clubs: Antarctic scientists. National Science Foundation Principal Investigators. South Pole voyagers. I think this has been the most noteworthy achievement of my career to date, and I feel very blessed to have had this experience. (2018)

His enthusiasm is humble, and his love for the place and the people is clear. His words highlight the rarity of his experience, but also illustrate the happiness and excitement with which many people write about the Ice.

For such a harsh continent, Antarctica brings a lot of joy.

**Kelly E. Hall** completed an MA in medieval studies at the University of York (UK) and a PhD in medieval literature at Florida State University. She is currently the Director of Global Initiatives at Cedar Crest College and directs their cornerstone programme, the Sophomore Expedition. She worked four austral summer seasons at McMurdo Station with the United States Antarctic Program, first as a vehicle operator and later as a logistics manager for the Shackleton Field Camp.

## Notes

1. Argentina's Esperanza Station is an exception. Families live there and when children are born, they receive Antarctic passports. Other countries who are members of the Antarctic Treaty do not legitimize this claim to citizenship.
2. Joe Kendall, a vehicle operator at McMurdo Station, celebrated his eightieth birthday on station during the 2009 season.
3. Despite the sometimes-harsh conditions, US workers do not make significantly more money than their non-Antarctic counterparts. As an example, I made USD 8.33 per hour driving vehicles during my first season in 2008, but according to the Bureau of Labor Statistics the average bus driver in America was making USD 15.61 per hour at

that time. In addition, most Antarctic workers are only there for temporary contracts that range from four to nine months. Antarctic employment comes with free travel, room and board, but most positions do not offer pensions or health insurance.

4. A different organization was selected each year, but it was usually a charity that supported women and/or children.
5. The Facebook page *Ice Friends Helping Ice Friends* is a group of over 2,700 current and previous workers. It was created in 2013 when a looming US government shutdown began to affect those heading to Antarctica, leaving many temporarily stranded or homeless as they learned their jobs had been cut. Since then, the page has become a solid resource for those looking for Ice work, those who are advertising work elsewhere, those seeking house or pet sitters or those seeking general travel advice. The question had received forty-four responses as of 1 July 2022.
6. For example, both writers visit Argentina's Esperanza Station and discuss the unique environment where families live together, and the station does not attempt to disguise their efforts at colonization. See Wheeler (1998: 268–69) and Walker (2021: 271–77).

## References

Bradford, Alesha, and Jarryd Salem. 2022. 'Science and Vodka at Vernadsky Station in Antarctica', *Nomadasaurus* (blog), 21 January. Retrieved 19 July 2024 from https://www.nomadasaurus.com/science-vodka-at-vernadsky-research-station-in-antarctica/.

Carlstrom, John E. 2002. 'Cosmic Microwave Background Stage 4', Department of Astronomy and Astrophysics, University of Chicago.

Dave and Deb (no last name given). 2021. 'A Zodiac Tour of Pleneau Bay, Antarctica', *The Planet D* (blog), 22 December. Retrieved 19 July 2024 from https://theplanetd.com/a-zodiac-tour-of-pleneau-bay-antarctica/.

Davidson, Matt. 2008. 'Suffer', in *Freeze Frames*. Morrisville, NC: Lulu Press, p. 52.

Drake, Neill. 2021. 'Don't Fear the Drake Passage – Embrace It', *Antarctica Travel Group* (blog), 22 February. Retrieved 19 July 2024 from https://www.antarcticatravelgroup.com/blogs/news/dont-fear-the-drake-passage-embrace-it.

Endo, Michelle. 2022. 'How I Got Paid to Live in Antarctica: FAQ about Working on the Ice', *Wandereatwrite* (blog), 21 September. Retrieved 19 July 2024 from https://wandereatwrite.com/how-i-got-paid-to-live-in-antarctica/#.

Haugen, Lori Johnson. 2022. 'On Our Last Deployment to Antarctica . . .', Facebook, 30 June. Retrieved 19 July 2024 from https://www.facebook.com/lori.haugen.12.

Huang, Nellie. 2012. 'Kodak Moment: Cruising through Lemaire Channel in Antarctica', *Wild Junket* (blog), 11 December. Retrieved 19 July 2024 from https://www.wildjunket.com/kodak-moment-cruising-through-lemaire-channel-in-antarctica/.

Jerrard, Megan. 2017. 'The Trip of a Lifetime: To Antarctica on an Expedition Cruise', *Mapping Megan* (blog), 27 March. Retrieved 19 July 2024 from https://www.mappingmegan.com/antarctica-on-an-expedition-cruise/.

Johnson, Nicholas. 2005. *Big Dead Place: Inside the Strange and Menacing World of Antarctica*. Los Angeles: Feral House.

Kiell (no last name given). 2022. 'Working in Antarctica – A Blogger's Experience', *Emily Luxton Travels* (blog), 26 April. Retrieved 19 July 2024 from https://www.emilyluxton.co.uk/working-in-antarctica/.

Kylie and Scott (no last name given). 2022. *Live Hard, Travel Often* (blog). Retrieved 19 July 2024 from https://lovehardtraveloften.com/?s=how+to+visit+antarctica.

Lipuma, Lauren. 2022. '"The Nutcracker" Comes to McMurdo', *Antarctic Sun*, 9 May. Retrieved 19 July 2024 from https://antarcticsun.usap.gov/features/4722.

McClanahan, Paige. 2020. 'Tourism in Antarctica: Edging toward the (Risky) Mainstream', *New York Times*, 26 February. Retrieved 19 July 2024 from https://www.nytimes.com/2020/02/26/travel/antarctica-tourism-environment-safety.html.

McCulley, Kate. 2018. 'Antarctica and the Traveller's Ego', *Adventurous Kate: Travel for Independent Women* (blog). Retrieved 19 July 2024 from https://www.adventurouskate.com/antarctica-and-the- travellers-ego/.

Nayak, Michael. 2018. *Antarctica 2018* (blog). Retrieved 19 July 2024 from https://www.michaelnayak.com/blog.

Nelson, Keri. 2022. 'Live from Palmer Station with Keri Nelson' (video), YouTube, 22 February 2022, 44:29. Retrieved 19 July 2024 from https://youtu.be/XODeSDhk-dk.

Ott, Sherry. 2022. 'How to Travel to Antarctica: The Ultimate Travel Guide', *Ottsworld* (blog), 18 May. Retrieved 19 July 2024 from https://www.ottsworld.com/blogs/how-to-cruise-to-antarctica/.

Schieffelin, Dan. 2021. 'Ah the Good Old Days. . .', Facebook, 15 August. Retrieved 19 July 2024 from https://www.facebook.com/dan.schieffelin.7.

Scotia, Jennifer. 2011. 'Best Boondoggle Ever!', *Jen and Scott on the Ice* (blog), 27 January. Retrieved 19 July 2024 from https://jenandscottonice.blogspot.com/2011/01/best-boondoggle-ever.html.

———. 2021. 'We Had a Partial Eclipse this Weekend', Facebook, 5 December. Retrieved 19 July 2024 from https://www.facebook.com/jennifer.scotia

Shackleton, Sir Ernest. 2000. *South! The Last Antarctic Expedition of Shackleton*. Santa Barbara, CA: Narrative Press.

Tweed, Michelle Majerus. 2017. 'This Was Fun to See. . .', post in *I've Been to Antarctica* [Facebook group]. Facebook, 17 September. Retrieved 19 July 2024 from https://www.facebook.com/groups/2212798205.

Walker, Gabrielle. 2012. *Antarctica: An Intimate Portrait of a Mysterious Continent*. Boston: Houghton Mifflin.

Wheeler, Sara. 1998. *Terra Incognita: Travels in Antarctica*. New York: Random House.

Windsor, Steven. 2021. 'Go on a Snowmobile Trip. . .', Facebook, 31 December. Retrieved 19 July 2024 from https://www.facebook.com/steven.windsor.5.

## CHAPTER 8

# Moving between Modes

## *Robert Macfarlane's and Kathleen Jamie's Journeys on Foot and in Time*

Monika Kocot

### Introduction

This chapter discusses journeys on foot and in time in Robert Macfarlane's *Underland* and Kathleen Jamie's *Surfacing*. Both books were published in 2019, and both touch upon the issue of 'deep time' and our (human) connection to the past and possible future(s) of our planet. What makes them strikingly different – and what this chapter will mainly dwell upon – is the way Jamie and Macfarlane approach the issue of the journey itself in terms of mode(s) of travelling, and how they explore the relationship between the traveller and the land under their feet. Both authors are keen practitioners of what Kenneth White calls 'voyage-voyance' (travelling-seeing – learning how to look in order to see). I am referring to White's concept, as in my view it perfectly matches Macfarlane's and Jamie's travel narratives, their attention to detail, their openness to whatever circumstances they encounter on their way. Even though their approach to writing is different, White's theory of 'travelling-seeing' might shed new light on what each writer decides to foreground in their writing practices.[1]

### Voyage-Voyance/Travelling-Seeing?

In an essay titled 'Petit album nomade', Kenneth White, a famous essayist, writer-traveller and the founding father of the geopoetic project, explores the philosophy of his transcendental travelogues and dwells on the way his travels (*voyages*) are related to two words in French that are near-homophones: *voie* (way) and *voir* (seeing):

> In a series of books, *La figure du dehors*, *Une apocalypse tranquille*, *L'espritnomade*, I tried (these are indeed essays [attempts], an explor-

> atory genre) to clear the space of that literature and to constitute a genealogy of a kind of travel narrative belonging to a new type, which can of course take several forms, assume several tones, according to the personality of the writers. In English, I called it waybook, in German *Wegbuch*, and in French, for lack of a similarly short term: *voyage-voyance*. In all these terms, in addition to the notion of travel, there is the notion of a path (the life line) and of seeing (perceiving another space, opening other dimensions). If this other space, other dimension, is not there, one does not escape the 'universal reporting' that was stigmatized already by Mallarmé. (White 1999: 180)

Obviously, the name *voyage-voyance* reveals much more than *Wegbuch* or waybook. As I have argued elsewhere, it is important to note that in White's vision the notion of the way (*voie*) is inextricably linked with the line of life (*ligne de vie*) and with seeing (perceiving that other space, opening other dimensions) (Kocot 2019: 113; 2020: 55; 2023: 121). Without the element of vision and opening our eyes to what appears on our path, we do not speak about the waybook but about 'universal reporting'. As Mohammed Hashas aptly notes in his book on White's geopoetics, we need to see travel (*voyage*) as both mental and physical 'so that the mindscape will correspond to the landscape, and vice versa, which will in turn be expressed in a particular wordscape' (2017: 49). In my view, Jamie's and Macfarlane's travel narratives are prime examples of the interplay of mindscapes, landscapes and wordscapes, and I will attempt to show how these dynamics operate in Jamie's *Surfacing* and Macfarlane's *Underland*.

Here I would like to offer a brief comment on the differences between these two collections of essays in terms of mindscapes. I am at one with William Dalrymple, who in his review of *Underland* writes that apart from 'occasional moments of epiphany and even ecstasy', Macfarlane's 'journeys deep into the earth "far from the human realm", are usually melancholic and claustrophobic, and are occasionally properly frightening' (2019). Dalrymple sees the connection between *Underland* and Macfarlane's first book, *Mountains of the Mind*, where obsession with fear and awe-inspiring landscapes is often accompanied with savouring the adrenaline rush. For instance, under Paris, Macfarlane nearly becomes stuck in a narrow vertical shaft: 'How long does this tunnel run like this? If it dips even two inches, I'll be stuck. The thought of continuing is atrocious. The thought of reversing is even worse. Then the top of my head bumps against something soft . . . I can feel the stone around me, the stone that encases me, the stone that is measuring me up like a coffin, *starting to vibrate*' (2019: 168). In the story 'Descending', which opens Part 1 of the book, Macfarlane writes: 'I have rarely felt as far from the human realm as when only ten yards below it, caught in the shining jaws of a limestone bedding plane first formed on the floor of an ancient sea' (ibid.: 11). To be 'caught in the shining jaws of a limestone bedding plane' is precisely something Kathleen Jamie would like to avoid. It is not a secret that she has criticized Macfarlane's travel writing – specifically in a review of his second book, *The Wild Places* (2007) – as the work of a 'lone enraptured male' (Jamie 2008), by 'highlighting the construction of wildness on which it depends' (Forsdick 2016: 73). And even in *Surfacing*'s opening essay, in which Jamie explores the entrance to a mountain cave, she offers an indirect (trickster) comment on Macfarlane's way of writing. She sets the scene first:

> You're in 'Reindeer Cave,' where antlers rather than bone were discovered. An excavation in the 1920s produced hundreds of reindeer antler, almost all from females.
>
> You sit at the cave-mouth, looking out at the rain, thinking about the Ice Age.
>
> You realise you haven't a clue. We can wait, say the hills. Take your time. (Jamie 2019: 1–2)

There is no adrenaline rush, no excitement whatsoever. What's truly exciting is that by means of minimalist gestures Jamie makes us readers enter the deep time journey. Let me come back for a moment to Jamie's review of *The Wild Places*, which, in her opinion, contains 'an awful lot of "I"', the result being that we don't behold the wild places, but the author (Jamie 2008). Jamie has made it clear that a lot of 'I' can be annoying even in her own writing: 'I annoy myself . . . I look at a page I've written, see that I've used the word 'I' 17 times and go back and reduce it by two thirds' (Jamie 2020). By applying that simple technique, she minimizes her presence in the narrative and makes more room for the wild or for the reader. In 'Reindeer Cave', Jamie writes 'you' in order for the readers to feel this space as their own, but also to foreground the issue of a larger community of beings taking part in the continuous process of creation. 'Talking' hills, witnessing human struggle to 'understand', to make sense of it all, make me think of Gary Snyder's *Practice of the Wild* and his chapter on Dōgen's 'Mountain and Water Sūtra' (1990). One can notice a similar attitude of humility towards nature, openness to mystery, but also the sense of letting go of ego control. This is obviously connected to viewing nature – and the wild – not as something external but both external and internal, or neither external nor internal – beyond duality. As Deborah Lilley notes:

> Consideration of what is encompassed by the term 'nature' recurs throughout her [Jamie's] writing from different perspectives, and in the process, the category of 'nature writing' is reconfigured to accommodate those perspectives. Her versions of the form are composed

> from ecological, social and historical viewpoints, challenging both the idealisation of nature and the perceived externality of nature that enables such idealisation. (2013: 16)

Jamie is not interested in the work of cave-divers: 'It makes you quail', she writes, 'the thought of crawling through darkness and passageways and underground streams. Echoes and falling rocks' (2019: 2). Interestingly enough, this is what Macfarlane dwells on in his *Underland*. Jamie is more interested in writing about memory. In her description of antlers stored at the National Museum of Scotland, she focuses on seemingly insignificant details and she interweaves the strands of deep time story and memory, including animal memory: 'Ancient fragments, they look more like broken biscuits. Also in the store, wrapped in a box, are the bear bones, including the brown-stained skull. The skull was in the cave and what was in the skull? Bear mind, bear memory' (ibid.: 3). This subtle gesture reminds us that human beings are not (and have never been) at the centre of the history of Gaia. We are only visitors, Jamie seems to be saying. She is interested in exploring the 'spirit' of a place, and that includes all the inhabitants of a given place. As Gary Snyder puts it in his *Practice of the Wild*, 'to know the spirit of a place is to realize that you are a part of a part and that the whole is made up of parts, each of which is whole. You start with the part you are whole in' (1990: 41).

In her review of Macfarlane's *The Wild Places*, Jamie offers an insight about things she considers wild: 'A dandelion poking up between paving slabs is natural and wild – and cheeky and subversive – but it doesn't carry that special wide-eyed sense of "wild"' (2008). One might argue that Jamie's mindscapes-wordscapes carry much lighter tones than those in *Underland*; here the focus is more on the bright – and cheeky and subversive – side of (re)surfacing things, even if at first sight they might connote something dark or frightening. The difference lies also in the way she approaches her themes and the reader. As Dani Garavelli puts it,

> *Surfacing* is about excavation, literal and metaphorical. The two longest essays describe sites in a Yup'ik village in Alaska and Links of Noltland, Orkney. She is both archivist and archaeologist, unearthing artefacts and memories, digging, as Seamus Heaney would have it, with her pen; but also scattering new fragments, like broken china, for future generations to sift through and interpret (2019).

Jamie operates by means of understatement, and she does not posit herself at the centre of the scene; there is room for the readers to complete the image with their own mindscapes, if you will. If Macfarlane operates by means of evoking intense emotions and by building suspense, Jamie captivates readers using subtle modes of poetic diction. In his review of *Surfacing*, Tim Hannigan notes that her writing 'has the meticulous precision of poetry' and he adds that 'its fineness has an elusive, quicksilver quality' (2019). Indeed, the way Jamie orchestrates the laws of constant change and impermanence can be surprising, even for readers acquainted with her previous books.

In a way, even the titles of the books in question might give us an insight into the nature of these travel narratives. Macfarlane chooses to search for the unknown by going under the surface of the Earth, by exploring the mysteries hidden there; Jamie does the opposite – she mindfully notices what comes up to the surface, and usually begins her stories with a seemingly insignificant detail that enters the (landscape) scene of its own accord, as if by coincidence. A comparative reading of *Surfacing* and *Underland* might become a fascinating exploration into each author's frame of mind and their practice of 'framing' the landscape into wordscape. And White's insights as a poet-traveller can be helpful in opening more vistas within Macfarlane's and Jamie's writing.

## On the Art of Asking the Right Questions

In *L'Ermitage des brumes: Occident, Orient et au-delà*, White notes that apart from being travel accounts, waybooks are books of the way (*livre(s) de la voie*); the focus is placed on spiritual affirmation of the new, opening our eyes to the unknown (Kocot 2019: 113; 2020: 55):

> While traversing territories, horizontally, if I may say so, these books try to discover cultural paths overshadowed by history, ways of thought (a place, a moment, perhaps an opportunity, not for a vague reflection but a turning point of the spirit), and pathways of feeling, where the freshest possible sensations appear: 'each step pure wind', as the zen koan says. Within the journey [*voyage*], there is always the way [*voie*] (waybook – a book of the way, not only a travel book) – but in a discreet way. (White 2005: 79)

In *Le poète cosmographe*, White argues that travelling (*voyage*) should be accompanied with *voyance*, 'with knowing how to look in order to see (*savoir voir*), with mindful being in the world, and with constant movement' (Kocot 2019: 113; 2020: 56):

> There are many things to discover, starting with the slightest sign, about Pict culture, Celtic culture, the life of the Vikings .... For that, of course, one needs to be on the alert. Travelling for me is about moving, of course, and I like movement, but it's also about knowing how

to see. That's why I'm talking about sites-sights [*voyages-voyances*]. (White 1987: 53)

That travelling is knowing how to see is clear. But to see what, one might ask. In my opinion, both Jamie and Macfarlane try to notice what is happening to our planet in the era of the Anthropocene, but the feeling they get while travelling-seeing is often different in terms of emotional intensity, emotional temperature, if you will. Jamie is much more grounded in the here and now, in the nowness of what is happening, and if she goes back to the past she does so in order to revisit cultural narratives. For instance, in a story titled 'Reflection', Jamie applies a postcolonial perspective and connects the story of the False Bride from an old ballad, an account by William Scoresby about a reflection of a ship in the sky and the practice of cultural appropriation. Her focus is on one of the items in the Arbuthnot Museum in Peterhead:

> These Inuit things had been bartered for, the curator had said. 'Bartered for what?' I remember asking.
> 'Guns! And there were a lot of bairns born there, to Inuit women and brought up there, and the wives at home had no idea!'
> False husbands, warmed by their icy fire. (Jamie 2019: 10)

One can also notice a sense of acceptance and letting go, as in a story about picking up fragments of some old crockery from the ploughed field: 'They fill your hands, these fragments, these stories, but with a wide gesture, you cast them back across the field again' (ibid.: 177). When Jamie wishes to signal her concern about the future of our planet, she writes one sentence, maybe two, and then she leaves room for the reader to draw conclusions. Her writing is full of understatement, of ellipsis, of things unsaid. We can call these moments silences. The importance of irony in Jamie's work should also not be underestimated. As Marina Benjamin puts it, 'Every now and then . . . [Jamie] cuts through the assemblage of beautiful prose with a stinging comment; a reminder that the natural balance is out of whack, or that violence and menace can surface just as easily as venerable artefacts from the past' (2019). In the first story, 'The Reindeer Cave', we get a mixture of ellipsis and irony:

> There were arguments about land management, flood-plains, deforestation. Commentators intoned, 'Is it climate change?'
> Well, you thought sourly, if it walks like a duck and quacks like a duck, it's a duck.
> *D'où venons-nous ? Que sommes-nous ? Où allons-nous ?* (Jamie 2019: 4)

In the last story, 'Voice of the Wood', Jamie presents an interesting take on the idea of being lost, on being anxious, which might be treated as one more subtle gesture towards deconstructing the narrative of 'a lone enraptured male' (Jamie 2008):

> How lost is lost?
>
> You are not lost. You followed your map. There is a path – there is always a path through the wood; there has been since the dawn of time . . . . Voice of the forest . . . Did you hear something move out the corner of your eye? The same moth come back? Or another leaf falling? You are not lost, just melodramatic. The path is at your feet, see? Now carry on. (Jamie 2019: 243, 245)

Macfarlane is much more interested in presenting a grand scale of things, with mythical, literary, anthropological, psychological references, with quotations from and references to other nature and travel writers. The idea of mapping the problem or weaving a story is of utmost importance here. In an interview with Garnette Cadogan (2020), Macfarlane admits that he is generally more interested in the pattern than the plot, and that his writing is a result of his obsessions. One of his 'obsessions' is a deep sense of solastalgia he experiences during his travels around the world. Solastalgia is a term coined in 2003 by the Australian philosopher Glenn A. Albrecht to mean 'the lived experience of distressing, negative environmental change' (2019: x). As Macfarlane puts it in his article 'Generation Anthropocene: How Humans Have Altered the Planet for Ever', '[w]here the pain of nostalgia arises from moving away, the pain of solastalgia arises from staying put. Where the pain of nostalgia can be mitigated by return, the pain of solastalgia tends to be irreversible' (2016). Near the end of *Underland*, in the chapter 'The Hiding Place in Finland', we find the following questions: 'what legacies will we leave behind, not only for the generations that succeed us but also for the epochs and species that will come after ours? Are we being good ancestors . . . ?' (Macfarlane 2019: 410). And here's one more sentence with references to the *Kalevala*: 'maybe this is among the best things we can try to do, I think, as the Sampo grinds through the world's epochs: to be good ancestors' (ibid.: 419).

Jamie and Macfarlane may well be asking similar questions concerning the age of the Anthropocene, but their mindscapes and wordscapes are different. Instead of lofty questions about being 'good ancestors', Jamie opts for short messages on the politics of climate change and ecological threat that make the readers stop and think. In 'Links of Noltland I', she offers a series of questions that are definitely less abstract than those in *Underland*:

> Does this matter, is the question. Do we want to know how it was to be human, here, five thousand years ago? Do we want to know where we're coming from as we cruise into the future? What we were, or might be again? How we 'engaged', if that's the word, how we configured our relationship with the rest of the natural world, with the planet. (2019: 136)

The difference of perspective of looking at climate change is visible also on the covers of *Underland* and *Surfacing* (first and second editions). The first cover of *Surfacing* (with a photo by Erika Larsen) is minimalistic, evoking the sense of looking at objects that are brought to the surface of the Earth (a few stones, bones – perhaps pieces of jewellery – fragments of ancient arrows) or water (a feather, a leaf, seaweed). The arrangement of objects on the page makes one think of a neat display of things found at an excavation site. The second cover (photo by Wayde Carroll) is more about the idea of water (the pond?) reflecting things (the sky, the landscape). It is interesting to note that we approach the scene at the level of leaves of grass; they occupy the foreground, while the background is all sky, doubled as it were by its reflection in the water. One cannot help thinking that the dark elements in the picture are here for a reason, as in Jamie's writing darkness, be it literal or metaphorical, is also a medium of vision.

The cover of Macfarlane's *Underland* features a painting titled *Nether* by Stanley Donwood, Macfarlane's long-time friend and collaborator. As Macfarlane puts it in the last section of acknowledgements:

> while Nether looks like a vast sun rising at the end of a sunken lane, it's not. I remember asking Stanley about the image ... 'Nether', Stanley said then, 'isn't the sun. It's the last thing you'd ever see. It's the light of a nuclear blast that has just detonated, seen down a holloway. When you look at Nether, you've got about 0.001 of a second of life remaining, before the flesh is melted from your bones.' (2019: 468)

Solastalgia and 'the shadow of an annihilation', to use Sebald's phrase, is certainly there, in each line. But Macfarlane adds one more thing: 'Lustrous and lethal, fatal and beautiful, nuclear and natural, the image beckons the viewer's eye on and into the underworld, into its reactor core. As such, it could hardly be truer to the atmospheres of *Underland*' (ibid.). And that takes us back to Jamie's insight concerning Macfarlane's immanent poetics, his obsession with beauty and danger, a sort of addiction to focusing on the 'lustrous and lethal' faces of the world. One cannot think of a better contrast between this way of looking at the world and Jamie's affirmation and embrace of the commonplace, and the mystery of the world 'hidden' in plain sight.

## Watching the Surface of Things versus Going 'Under' the Surface

Just one look at these two collections of essays shows us that they were not and could not have been written in the moment of travelling. It is obvious that we are speaking of a long, post-travel editorial process. *Surfacing* is full of travel narratives dealing with loss, dealing with time, deep time, spiral time. Some of the stories explore Jamie's recent journeys to Alaska or Orkney; others, such as 'A Tibetan Dog' or 'Wind Horse' write through or revisit a period of travel previously recorded in her first prose book, *The Golden Peak* (1992; republished in 2002 as *Among Muslims*). Jamie brilliantly shows the connection between places and landscapes and personal memories and mindscapes. The use of phrases such as 'to surface', 'to resurface', 'to come to the surface', 'to be brought to the surface', 'to surface in one's mind' might give an idea of finding those extraordinary moments in ordinary situations, of being open to receiving these moments without much effort (Kocot 2023: 118). Each essay opens with a selected picture. In longer essays we may find a number of pictures taken by Jamie herself or by professional photographers. Macfarlane also opens each chapter with a photo, illustration (Chapter 9) or engraving (Chapter 6), and we can read about each of them in the acknowledgements (see Chapter 1 of this collection for a discussion of this deployment of images). But, as I have already mentioned, he is more interested in the pattern, the ornamentation. We can see that at the beginning of the first chapter: 'The way into the underland is through the riven trunk of an old ash tree' (2019: 3), and at the beginning of the last one: 'The way out of the underland is where nine springs flow clear from the bedrock' (ibid.: 423). More than that, at the end of the first chapter we find the following passage. Here even the typography shows new patterns:

> So these scenes from the underland unfold along the walls of this impossible chamber, down in the labyrinth beneath the riven ash. The same three tasks recur across cultures and epochs: to shelter what is precious, to yield what is valuable, and to dispose of what is harmful.
>
> Shelter (memories, precious matter, messages, fragile lives).
>
> Yield (information, wealth, metaphors, minerals, visions).
>
> Dispose (waste, trauma, poison, secrets).
>
> Into the underland we have long placed that which we fear and wish to lose, and that which we love and wish to save. (2019: 8)

We will not find such classifications or neat categorizations and labels imposed on the nature of our world in Jamie's book. We will not find any bibliography there, lists of quotations, notes or references. There is no division into parts (Macfarlane opts for three parts, headed 'Seeing', 'Hiding' and 'Haunting').

It seems that for Macfarlane the idea of exchange of ideas and theories is the most important. It becomes quite explicit in the penultimate chapter, entitled 'The Hiding Place', where we find the following passage:

> I think of the black and red hand-prints left on the cave walls at Chauvet, of the red figures of the dancers with their outstretched arms, of the spray-can hand stencil on the catacomb wall in Paris, of Helen reaching a hand down to haul me out of the moulin. I think of the many people I have encountered in and through the underland who have been committed to shared human work rather than retreat and isolation. Many of them have been mappers, really, of networks of mutual relation, endeavouring to stitch their thinking into unfamiliar scales of time and space, seeking not the scattered jewels of personal epiphany but rather to enlarge the possible means by which people might move and think together across landscapes, in responsible knowledge of deep past, deep future and the inhuman earth. (2019: 418)

He makes it perfectly clear that he is not interested in seeking 'the scattered jewels of personal epiphany', which, as I have already noted, is the very thing Jamie finds so exhilarating and mind-opening; for Jamie, those 'jewels' might offer a sort of gateway into the land of insights for other people, but she does not feel entitled to impose her way of thinking on her readers. On the contrary, she is fully aware of the role of our individual agency as an important element behind the change we would like to welcome.

There is an interesting passage in Macfarlane's essay 'The Hiding Place' where he quotes from *After Nature: A Politics for the Anthropocene* by Jedediah Purdy:

> People are best able to change their ways when they find two things at once in nature: something to fear, a threat they must avoid, and also something to love, a quality . . . which they can do their best to honour. Either impulse can stay the human hand, but the first stops it just short of being burnt or broken. The second keeps the hand poised, extended in greeting or in an offer of peace. This gesture is the beginning of collaboration, among people but beyond us, in building our next home. (2019: 419)

Commenting on this quotation, Macfarlane once again offers two options, and this bipolar thinking will resurface in many of his essays, usually along with a motivational message for his readers. Jamie's book does not offer any lofty message for humanity, and even if her essays do suggest some action on the part of the reader, they are very subtle, just like the framework of the whole book.

Jamie's structures are playfully rhizomatic (Deleuze and Guattari 1987: 3–28), embracing non-linear, decentralized modes of cognition, based on associative thinking, unpredictable. Understatement works miracles.

## Writing and Travel Experience

It is interesting to note that the post-editorial process in Macfarlane's case took six years, but strangely enough the idea for a book came much earlier than the actual travel. In one of the public readings at Politics and Prose (2019) he shares the story. We learn that in 2010 (nine years before publication) Macfarlane was moved by three events, three underland stories:

- In April 2010 the Deepwater Horizon oil spill (also called the Gulf of Mexico oil spill, the largest marine oil spill in history) occurred.
- A volcano, situated on the Eastern Volcanic Zone in southern Iceland, began erupting lava on 20 March 2010. A second eruption then began beneath the ice cap near the summit of the volcano on 14 April. This eruption caused the melting of large amounts of ice, leading to flooding in southern Iceland.
- The 2010 Copiapó mining accident, also known then as the 'Chilean mining accident', began on Thursday 5 August 2010, with a cave-in at the San Jose copper-gold mine, located in the Atacama Desert in northern Chile. Thirty-three men, trapped 700 meters underground and five kilometres from the mine's entrance via spiralling underground ramps, were rescued after sixty-nine days.

Macfarlane makes it clear that real-life events inspired him to write a book on the importance of the underland – not only for our past but also for our future. One could argue that the solastalgic element must have been the key force behind writing *Underland*.

In Jamie's collection, each essay tells a different story, and the whole collection is unified by means of subtle interlinking ideas and motifs. One of them – in my opinion certainly the most significant, and at the same time latent in terms of narration – is the nature of body-mind and how it connects with nature understood as the wild. In her insightful essay entitled 'Into the Centre of Things: Poetic Travel Narratives in the Work of Kathleen Jamie and Nan Shepherd', Eleanor Bell notes that for Jamie 'the mixture of mediating the natural world and the wild through the body, domesticating the wild in small, ordinary ways in order to make it at least partially comprehensible, can be regarded as central to her vision' (2015: 132). Even though Bell wrote that essay a few years before the publication of *Surfacing*, her observation still holds true, particularly in the story entitled 'A Tibetan Dog' in which the motives of travelling, writing (also taking notes), bodily sensations and memory become 'entangled' with the concept of healing.

The story in question opens with two photographs: in the first one, we can see a notebook with a picture of Buddha Shakyamuni, while the second shows Jamie sitting on the steps with a Tibetan novice monk – Jamie points her finger at something in the monk's notebook, and the monk looks at her with intense interest. In my article entitled 'Liminality and In-Betweenness in Kathleen Jamie's *Surfacing*' (2023), I argue that it is not clear who asks questions, and who provides answers, and that perhaps it is a scene of mutual exchange of ideas (Kocot 2023: 122). In 'The Wind Horse', Jamie gives more details regarding the scene on the steps. For her, it is a moment of connection:

> There were moments of connection, too. One evening I was walking by the monastery when a boy monk called to me, a novice of ten or twelve, with a sweet manner. He was sitting on some steps poring over a slate of schoolwork. What luck! On his slate was the English alphabet, and lo, here was a native speaker strolling by. He beckoned me over. He was doing well, could recite the letters, but he was defeated by 'W'. He pointed to the 'W' on his slate, and made a helpless gesture. I sat beside him, and a small group of shaven-headed boys in robes gathered to watch and laugh as I drew the letter carefully in my own notebook, then passed the pen. 'Dubble-you', I intoned. 'Dub-elle-you', the lad replied. We went on with a short cheerful lesson. I saw him and his class-mates several times after that. 'Dub-elle-you!' they'd call, waving across the street. (Jamie 2019: 211)

If, as Tim Hannigan puts it, 'the whole thing [collection] shimmers' and the more closely one looks, 'the more detailed its patterning becomes, with common threads and images running back and forth through the text' (2019), then by looking at 'A Tibetan Dog' from the point of view of the patterning one can really experience not only the spiral time multiverse (Kocot 2023: 120, 123), but also layers of meaning that are deliberately effaced in order to create more question marks for the reader. One needs to reread the story a few times in order to notice little 'gems' reflecting the messages that precede and follow given strands of narration. This story in particular foregrounds the theme of knowledge, or, better still, wisdom – and its counterpart, mystery, or the unknown. Because the story's time-space is cyclical, readers unravel subsequent folds of the mystery step by step. Given the description of the scene on the steps in 'The Wind Horse', one can see how much the perception of the photograph in 'A Tibetan Dog' changes; perhaps some of our projections concerning the scene without the background provided in 'The Wind Horse' are no longer valid, perhaps some new interpretations emerge. *Surfacing* does 'shimmer', and each time we change our perspective of looking at a given scene, all the elements of the story are seen in a new light. But it also means that some elements are no longer as clear and as foregrounded as they were before. Each piece of information affects our perception of the whole scene. The effect of this 'shimmering' extends well beyond a given story: It all begins like a fairy tale – 'Far away from here, in the province of Amdo in China, there lies a town called Xiahe' (Jamie 2019: 187) – but the second sentence immediately opens a political dimension: 'To the Tibetans, Amdo being a Tibetan province, the place is known as Labrang. It is the site of an important monastery' (ibid.). Near the end of the story we learn that the year is 1989. In 'The Wind Horse' we are given more details of the Labrang monastery, and the time of unrest in Lhasa, student protests in Beijing and in other Chinese cities. One might argue that the story is one of the most politically and spiritually engaged pieces in Jamie's writing. But apart from a political and ideological change taking place in this part of the world in 1989, 'A Tibetan Dog' speaks of a different kind of transition: from naivety and superficial contact with the autochthons to a more sophisticated (more open) mode of being in the world (Kocot 2023: 122–23). Quite surprisingly, the teaching comes through the body, and it results in awakening the awareness of body-mind and the sense of interbeing. The importance of travelling cannot be underestimated, as the whole process of linking very distant strands of narration is based on using the old travel notebook from 1989 and writing through the notes to arrive at the message hidden in the body twenty-five years later (ibid.: 123). Bodily sensation, which opens the narrative, is presented from the point of view of the young Jamie. The theme of anger and ignorance seems to dominate this part of the story:

> One evening many years ago, walking alone, I was about to cross that bridge from the Tibetan side when, emboldened by dusk, a little terrier came rushing out and nipped my calf. I didn't see it coming; the first I knew was the sensation: 'nip' is the very word. I felt a quick pain, so spun round to see the creature already scarpering back among the houses, even as I yelled. I think I threw a stone. The brute had rushed out specially to bite me, a foreigner. The tiny wound was dribbling blood, so back at the hotel I dabbed at it with iodine.
>
> I had two companions there: Sean, with whom I was travelling, and Elena, a young Italian woman who was staying down the corridor.
>
> 'Look,' I said. 'A dog bit me. A Tibetan one.'
>
> They didn't seem too bothered.
>
> 'They love dogs, the Tibetans,' said Elena. 'They say they are the reincarnations of lamas who . . . how do you say?'
>
> 'Didn't make the grade?' said Sean.

> 'Lama or no, the damn thing bit me.'
>
> And then the incident was forgotten. Of course it was. I was not much older than my own children are now. I had my travelling adventure, came home, a quarter-century passed. Partners were met and children were born and grew. Friendships were forged and lost. Jobs, projects, homes, bereavements, the stuff of life – if we're spared. The undammed rush of life.
>
> If we're spared. (Jamie 2019: 187–88)

It is interesting to see how seemingly unimportant themes – the significance of the bite as such, the role of the lama-dog, and Jamie's attitude towards the incident – become more and more meaningful as the story develops and envelops. By shifting the time frame, Jamie reveals a separate story that becomes curiously entangled with the first one. We now move in time to a not so distant past when Jamie underwent treatment for cancer: '[I]t was a week between the tests and the results. I already knew my lump was a cancer, but there were questions about how aggressive the cancer was, and how far it might have spread, and what treatments might be offered' (ibid.: 188–89). Jamie devotes one whole paragraph to discuss various emotional attitudes she 'tried on' and 'discarded' (ibid.: 189) like pieces of clothing none of which fitted, and she concludes by saying: 'why feel anything. It's all in the lap of the gods' (ibid.). In a way, Jamie suggests that by letting go of the 'right' kind of attitude, she invited the unexpected, the incomprehensible, the unknown in the form of a dream.

In the next part of the narrative the thin line distinguishing the world of wakefulness from the world of dreams becomes blurred, and Jamie is reassured by the vision in a dream. She orchestrates these strands in such a subtle way that the readers go through the narrative not even sensing that they are entering the mystical story of healing (Kocot 2023: 123–24).

> One night during that fitful week I managed to sleep some, and had a vivid dream. In the dream I was walking along a busy street when suddenly I felt a hard pinch on my left calf. Twisting round, I saw a small dog in the act of biting my leg. The dog waited till it made eye contact with me and only then, when it was sure I had seen it clearly, did it unclamp its jaws, release me, turn and trot away.
>
> What a strange dream! I woke with it clear in my mind, the bite, the dog's look. But something had changed in me. I woke relieved and strangely reassured that I wouldn't die of this cancer, not this time, not now. I was being nipped, and would be released. (Jamie 2019: 189).

As I have argued elsewhere, 'it is interesting to note that also the time frame of the here and now becomes "interwoven" with Jamie's past experiences in

Tibet' (Kocot 2023: 123). The dog from the past and the dream dog are one. But the scene of seeing the dog is not the most important here. What seems to dominate Jamie's dream vision is the aspect of feeling: 'The dream-nip was the very sensation I felt then, in the same place' (Jamie 2019: 189). The dog from the past clearly enters the here-and-now of the dream in order to bring the message of healing. Jamie obviously tries to rationalize the phenomenon: 'How funny, to think my subconscious must have waited till I'd fallen asleep, then gone rummaging through a million long-lost memories to find an image it could craft into a message I would wake from and understand. And what had it come up with, so triumphantly? That little dog, and its teeth' (ibid.: 189–90). But immediately after that she changes her tone and opts for three most striking sentences (at least for a 'practising' atheist):

> Now that it's all over, it pleases me to imagine that the lama-dog knew what it was doing back in 1989, that it was laying down an act of kindness, and that I'd regret the stone-throwing and cast a blessing after its memory instead. I'd thank it for becoming that dream-metaphor trawled up to reassure me in an hour of need.
>
> The dog is long dead now and quite gone, unless it is indeed caught in the eternal wheel of rebirths, in which case I wish it a happy incarnation, next time round. (Ibid.: 190)

What follows is the whole rhizomatic story of the dog, the memory of the bite and its message for the here-and-now at the hospital. We learn for instance that the dream 'had opened a hinterland of other memories' (ibid.), and that it 'reawakened' the experience of being in Tibet. It would be helpful to mention at this moment that in Tibetan Buddhism we speak of a *terma*, a 'hidden treasure', various forms of hidden teachings that often unravel at the moment when we are able to receive and unpack the teaching (Ricard 1997). To some extent, Jamie speaks of a similar situation – the bite works as a prophecy, it activates the potential of the future self already present in the past. The physical sensation (bodily memory) of the bite carried in the dream activates the link between travelling in the distant past, the suffering in the present and the healing potential for the future (Kocot 2023: 124). We might risk saying that the dream is more real than reality, because it offers an experience of true wakefulness and clear vision.

The experience of cleansing through a dream is inextricably linked with the process of writing about what happened in Tibet. Jamie mentions that even when the cancer was becoming a distant memory, the idea of writing down the story persisted: 'eventually I cleared my desk and cleared my mind as much as possible', she writes, 'the better to immerse myself and remember the town of Xiahe' (2019: 191). It took Jamie several weeks to come up with the story. The 'feeling' aspect is foregrounded once again (Kocot 2023: 124): 'Absorbed, I was able to carry the "feel" of those days through my present tasks, inhabiting two worlds at once, as writers often do' (Jamie 2019: 193). Another interesting issue is rereading old notes, and rewriting them. Jamie provides many details concerning the retrieved notebook with the image of the Buddha Shakyamuni. We learn that it is a postcard with a *thangka* (Tibetan Buddhist painting) Jamie bought in Tibet and pasted on the cover of her notebook. 'Seeing the card again, I dimly remembered buying it at a street stall, not knowing what it meant, just liking the exoticism of the image' (ibid.: 192). One should not underestimate the importance of not knowing, so characteristic of Jamie's writing, especially in this collection. But also the appreciation of each moment, in each phase of life, the interconnectedness of things or events:

> The handwritten notebook with its picture postcard on the front. Museum pieces, but they seemed like old friends, which they were, for what is a photograph or a notebook? A reaching out to a future self, the future self of next week or half a lifetime. I was the utterly unimagined future person the notes had been made for, although I didn't know it back then. (Ibid.)

The *thangka* presents the Miracle at Sravasti, one of the great miracles performed by the Buddha. There are two main versions of the story. The Pali account of the miracle can be found in the *Dhammapadattakatha* and the Sanskrit version in the *Pratiharya-sutra*. In her description of the piece, Jamie decides to emphasize the wisdom elements: 'The card shows Shakyamuni, with blue hair and long earlobes, sitting enrobed on a stylised lotus flower, almost floating above many fiery demons. Amid calm yet sumptuous shades of blue and orange, he represents perfected wisdom, all-compassion' (ibid.). Her mention of the *thangka* offers a latent insight concerning the non-obvious nature of reality and our perception of reality. In the *thangka*, 'sumptuous shades of blue and orange' symbolize water and fire. According to Buddhist scriptures, at first the Buddha emitted fire from the top half and streams of water from the bottom half of his body; then he began alternating the fire and water elements between the positions only to shoot them up to illuminate the cosmos while he taught the Dharma to the awestruck observers of the miracle. The perfected wisdom the Buddha represents seems to be linked with the practice of going beyond notions, with transcending opposites, with affirming the law of non-duality. In Jamie's story, two short sentences describing the *thangka* are followed by three sentences that activate another aspect of her bodily memory:

> The Buddha on the front cover and, when I opened it, the notebook released a particular faint smell: paper, of course, maybe ballpoint ink,

> but also a herby, dusty aroma. Yes, of course: I'd picked some wildflowers at the shore of Qinghai Lake and pressed them between its pages. They were still there, flower-ghosts, fit to crumble at the slightest touch. (Ibid.: 192–93)

In that last paragraph on the processuality of retrieving her sensual memory of her past experience, Jamie dwells on the sense of smell. 'A herby, dusty aroma' of the 'flower-ghosts' has the power of transporting her back in time. In the last section of her story, she reflects on the practice of writing about her past experiences, of 'carrying the "feel" of those days through' her present tasks (ibid.: 193), and in doing so, Jamie decides to foreground the issue of not knowing the answers to some questions: 'Are we supposed to feel something when we encounter our younger selves? Nostalgia? I felt no nostalgia. Are we supposed to feel something when told we have cancer? What? Says who?' (ibid.). It might be argued that at the moment we let go of the need to conform to some unwritten rules that are supposed to 'guide' our behaviour, our feelings, we open up the space for deep insights concerning our genuine needs. 'It was late winter as I wrote, the sky darkening the window in the afternoons. What emerged were questions, to which I still have no answer' (ibid.), Jamie writes. The issue of not knowing is precisely what links Jamie's past adventures in Tibet, her struggle with cancer twenty-five years later and her reflections on both of those transformative experiences at the time of writing the story itself. For Jamie, the lack of answers is liberating, and she gets this message across to her readers on a number of occasions, each time from a different perspective. I argue that this is associated with her way of travelling-seeing.

If we follow Jamie in her story beginning in Tibet, moving through cancer treatment and ending with dream-reality loop, it is not difficult to see the whole narrative as a peculiar meditation on the nature of reality (dream versus wakefulness), on the continual process (processuality) of going through transition and challenge, but also on the practice of equanimity (connected with the practice of non-attachment and letting go). In Sanskrit, equanimity is *upeksha*; *upe* means 'over' and *ksha* means 'to look', so *upeksha* means 'to look over', 'to see all the sides'. In a way, throughout the narrative Jamie foregrounds the idea and practice – ideas without practice are empty slogans – of seeing with patience and understanding, which might offer new 'hues' of equanimity, namely inclusiveness, which gives it a more dynamic quality, and compassion.

The idea of letting go of conceptual understanding is foregrounded once again in the last two paragraphs:

> I deliberately avoided the Internet, and its unnuanced, memory-crushing, what-actually-happened gobbets of 'information'. No dream-dogs there.

> Perfected wisdom, all-compassion. I wonder where Elena is now. (Jamie 2019: 193)

We might interpret the last paragraph as a reference to the *thangka* symbolizing 'perfected wisdom, and all-compassion'. As I have argued elsewhere, 'perhaps that is all we need to know' (Kocot 2023: 125). This is what Jamie's minimalist phrasing seems to suggest. Perhaps 'perfected wisdom, all-compassion' offers a non-obvious gateway into the realm of knowledge of who we really are. That other knowledge is clearly contrasted with the Western idea of 'what-actually-happened gobbets of "information"'.

## Conclusion

In their travel adventures, both Macfarlane and Jamie stress the importance of being in the present moment, of being as wakeful and mindful as possible. Taking full responsibility for our actions and working for the benefit of our planet are absolutely crucial for both authors. What makes their narratives different is the mode of travelling-seeing and the mindscapes. The desired state of equanimity, understood as avoiding the state of being melodramatic, which is so clear in Jamie's narratives, stands in huge contrast with Macfarlane's insatiable desire for new adventures, driven by a mixture of fear of the future and the need to learn more, to explore the depths of the unknown. By carefully foregrounding the mythological, literary, anthropological and psychological patterns of the problem, Macfarlane attempts to shed (new) light on the unknown. Jamie, on the other hand, playfully affirms the presence of the unknown, she consciously centres the plot of her stories on the hidden, the elusive, the imperceptible. In Jamie's writing, empty spaces (in the form of understatement) and the focus on the mystery of being (letting go of completing the image) speak louder than words, they open the space of the in-between where true knowledge begins. Macfarlane's *voyages-voyances* are centred on experiencing the joy and exhilaration of uncovering the mystery of the underland, of exploring the unknown, while Jamie's sites-sights, tinged with equanimity, foreground that which cannot be seen and whose agency can be clearly felt when one can be still, when one is ready to appreciate it once it comes to the surface.

## Acknowledgements

All extracts from *Surfacing* (Sort of Books, 2019) by Kathleen Jamie are reproduced by kind permission of Sort of Books.

**Monika Kocot**, PhD, is Assistant Professor in the Department of British Literature and Culture at the University of Łódź, Poland. She is the author of *Playing Games of Sense in Edwin Morgan's Writing* (2016) and co-editor of *Języki (pop)kultury w literaturze, mediach i filmie* (2015), *Nie tylko Ishiguro: Szkice o literaturze anglojęzycznej w Polsce* (2019) and *Moving betweeen Modes: Papers in Intersemiotic Translation; In Memoriam Professor Alina Kwiatkowska* (2020). She has published articles on contemporary Scottish poetry and prose, as well as Native American writing. She is a member of the Association for Cultural Studies, the Association for Scottish Literary Studies, the Scottish Centre for Geopoetics and the Polish Cognitive Linguistics Association. She is the Vice President of the K.K. Baczynski Literary Society.

## Note

1. This chapter draws on one of my previous publications on Kathleen Jamie, which itself emerged from work initially conducted as part of the *Writing on the Move* project.

## References

Albrecht, Glenn A. 2019. *Earth Emotions: New Words for a New World*. Ithaca, NY: Cornell University Press.

Bell, Eleanor. 2015. 'Into the Centre of Things: Poetic Travel Narratives in the Work of Kathleen Jamie and Nan Shepherd', in Rachel Falconer (ed.), *Kathleen Jamie: Essays and Poems on Her Work*. Edinburgh: Edinburgh University Press, pp. 126–34.

Benjamin, Marina. 2019. 'Kathleen Jamie's Surfacing: Documenting the Beauty of Our Endangered Earth', *New Statesman*, 4 September. Retrieved 16 October 2022 from https://www.newstatesman.com/kathleen-jamie-surfacing-climate-change-essays-review.

Cadogan, Garnette. 2020. 'P&P Live! Robert Macfarlane | UNDERLAND', YouTube, 25 November. Retrieved 22 October 2022 from https://www.youtube.com/watch?v=i2tlV1bv0R4.

Dalrymple, William. 2019. '*Underland* by Robert Macfarlane Review – A Dazzling Journey into Deep Time', *Guardian*, 8 May. Retrieved 16 October 2022 from https://www.theguardian.com/books/2019/may/08/underland-by-robert-macfarlane-review.

Deleuze, Gilles, and Félix Guattari. 1987 (1980). *A Thousand Plateaus*, trans. B. Massumi. Minneapolis: University of Minnesota Press.

Forsdick, Charles. 2016. 'Travel and the Body: Corporeality, Speed and Technology', in Carl Thompson (ed.), *The Routledge Companion to Travel Writing*. Abingdon: Routledge, pp. 68–77.

Garavelli, Dani. 2019. 'Kathleen Jamie: "I'm Not an Emotional Person"', *Herald*, 7 September. Retrieved 20 October 2022 from https://www.heraldscotland.com/life_style/arts_ents/17875856.kathleen-jamie-im-not-emotional-person/.

Hannigan, Tim. 2019. '*Surfacing* by Kathleen Jamie', *Asian Review of Books*, 10 September. Retrieved 16 October 2022 from https://asianreviewofbooks.com/content/surfacing-by-kathleen-jamie/.

Hashas, Mohammed. 2017. *Intercultural Geopoetics in Kenneth White's Open World*. Newcastle upon Tyne: Cambridge Scholars.

Jamie, Kathleen. 2008. 'A Lone Enraptured Male', *London Review of Books* 30(5): 25–27. Retrieved 18 October 2022 from https://www.lrb.co.uk/the-paper/v30/n05/kathleen-jamie/a-lone-enraptured-male.

———. 2020. 'Kathleen Jamie, Chitra Ramaswamy & Amanda Thomson:*Antlers of Water*', interview, Edinburgh International Book Festival, broadcast 24 August 2020. Retrieved 16 October 2022 from https://www.edbookfest.co.uk/the-festival/whats-on/kathleen-jamie-chitra-ramaswamy-amanda-thomson-antlers-of-water/player.

———. 2019. *Surfacing*. London: Sort of Books.

Kocot, Monika. 2019. 'Kenneth White's North Road Travelling', in Anna Suwalska-Kołecka and Jakub Ligor (eds), *W drodze/On the Road – Perspektywy badawcze*. Płock: Wydawnictwo Naukowe Mazowieckiej Uczelni Publicznej w Płocku, pp. 107–25.

———. 2020. 'Writing the Road: On Drifting and Travelling-Seeing in Kenneth White's Geopoetics', *ANGLICA: An International Journal of English Studies* 29(3): 45–62.

———. 2023. 'Liminality and In-Betweenness in Kathleen Jamie's *Surfacing*', *Studia Litteraria Universitatis Iagellonicae Cracoviensis* 18(2): 115–27.

Lilley, Deborah. 2013. 'Kathleen Jamie: Rethinking the Externality and Idealisation of Nature', *Green Letters: Studies in Ecocriticism* 17(1): 16–26.

Macfarlane, Robert. 2016. 'Generation Anthropocene: How Humans Have Altered the Planet for Ever', *Guardian*, 1 April. Retrieved 10 October 2022 from https://www.theguardian.com/books/2016/apr/01/generation-anthropocene-altered-planet-for-ever.

———. 2019. *Underland: A Deep Time Journey*. London: Hamish Hamilton.

———. 2019. Robert Macfarlane, "Underland", Politics and Prose, broadcast 27 July 2019 Retrieved 10 October 2022 from https://www.youtube.com/watch?v=aWsRPNFfpDY&t=652s.

Ricard, Matthieu. 1997. 'Teachings: The Nyingma Lineage', *Rabsel* 5, Shechen Publications. Retrieved 31 October 2022 from https://web.archive.org/web/20070103012347/http://www.shechen.org/teach_nyima_mat.html.

Snyder, Gary. 1990. *The Practice of the Wild: Essays by Gary Snyder*. Berkeley: Counterpoint.

White, Kenneth. 1987. *Le Poète cosmographe: Entretiens*, trans. J. Fruzińska. Bordeaux: Presses Universitaires de Bordeaux.

———. 1999. 'Petit album nomade', trans. J. Fruzińska, in Alain Borer et al. (eds), *Pour une littérature voyageuse*. Brussels: Editions Complexe, pp. 167–96.

———. 2005. *L'Ermitage des brumes: Occident, Orient et au-delà*, trans. J. Fruzińska. Paris: Éditions Dervy.

## CHAPTER 9

# 'Did They Even Go There?'

## *Virtual Experience and Projected Journeys in Contemporary Travel Guidebooks*

Tim Hannigan

### Introduction: Experiencing a Rhinoceros

This chapter seeks to problematize the obvious assumption that bodily movement on the part of the author is intrinsic to the practice of travel writing, specifically in its contemporary functional subset, the guidebook. This assumption has apparently held good for centuries, and has, indeed, served as a way to differentiate between bona fide travel writers and 'travel liars' (Adams 1962). The *writing* may take place in static conditions at a desk; but the data that informs it must be the product of the experience of *travel* – perhaps initially recorded in notes made on the move. A writer who has not, in fact, moved to and through the places they describe may be declared a fraud – with the fourteenth-century author of the fabricated *Travels of Sir John Mandeville* the obvious archetype here. When it comes to informational travel writing, however, such an assumption has always been an oversimplification. And when it comes to contemporary guidebook-writing practice it may be wholly inaccurate – not because of any recent epidemic of mendacity among guidebook writers, but because of the new complexities in the nature of experience introduced by information technology.

This chapter begins by considering the awkward status of guidebooks within the existing scholarship of travel writing. It then briefly discusses the practical conditions of contemporary guidebook production, before exploring the issues outlined above with reference to examples from my own previous career as a guidebook writer, examining the use of 'virtual travel' as a data source and considering the complex relationship between the 'suggested itineraries' commonly featured in guidebooks and the actual travel experience of both writer and guidebook user.

But first, to emphasize the intended problematization outlined above, I wish to set readers of this chapter a simple task: draw a picture of a rhinoceros.

It need only be a very basic sketch; and it should be completed quickly without any direct recourse to reference material. The technical standard of the drawing will depend very much on the reader's own technical ability, but my purpose here is not to test artistic aptitude. Rather, it is to draw attention to what informs the picture, and to how our own situation regarding what we might call 'rhino data' differs radically from that of another artist undertaking the same task five hundred years ago.

Albrecht Dürer's famous woodcut of an Indian rhinoceros, produced in 1515, is almost certainly far more technically accomplished than any effort produced by a reader of this chapter. But – varied ability notwithstanding – a contemporary effort will almost certainly be *aimed* at a more objectively accurate representation of the actuality of a rhino. It is quite possible that readers of this chapter have no more seen a rhino in the flesh than had Dürer; and yet, even so, they will have far more *experience* of rhinoceroses than was ever possible for a German artist in the sixteenth century. Dürer's image was based on a written description of a live rhino recently delivered to Portugal from India, probably accompanied by a basic eyewitness sketch, as well as Pliny the Elder's

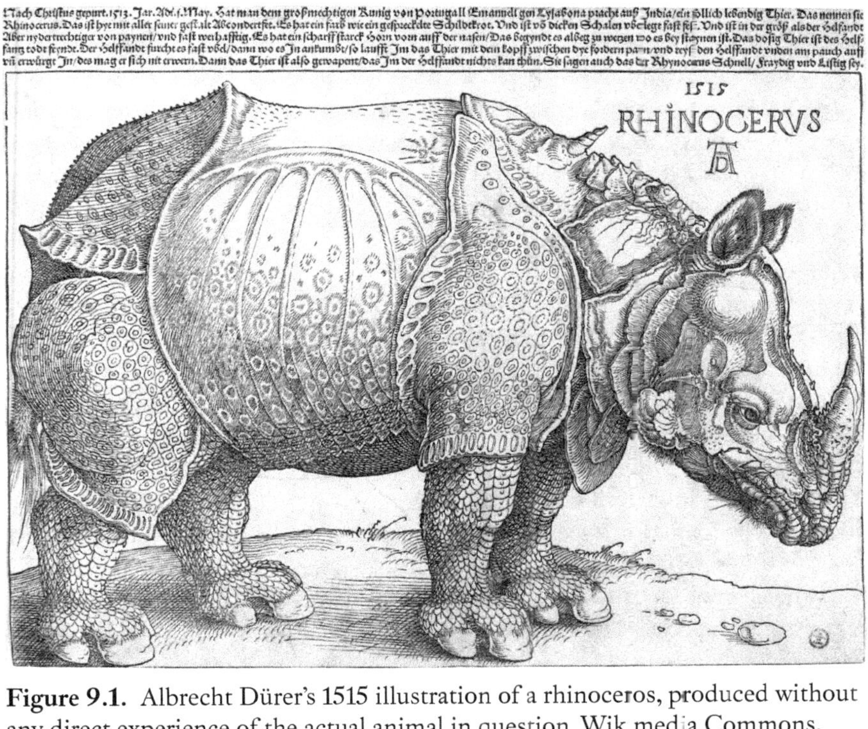


**Figure 9.1.** Albrecht Dürer's 1515 illustration of a rhinoceros, produced without any direct experience of the actual animal in question. Wikimedia Commons, public domain.

1,500-year-old literary account of the animal (Vertue 2020). Most readers of this chapter, meanwhile, will draw on a lifetime of exposure to photographic and filmic records of actual rhinos, as well as myriad more accurate drawings and paintings. Literary descriptions probably have minimal importance. And even if an individual reader has actually seen a real rhino, in a zoo or in the wild, that is probably less significant to their ability to accurately draw a rhino than the mass of rhino images to which they have been exposed since childhood. And conversely, having never directly encountered a rhino would not be considered a fundamental impediment to drawing a reasonably accurate picture of one.

This initial digression serves, I hope, to raise important questions about what it means to have access to empirical data, and what it means to have *experience* of that data, in the twenty-first century. All of this is pertinent to a question sometimes cynically asked of modern guidebook writers, as much as of Marco Polo and other medieval travellers to far-off lands: *did they even go there*?

## Guidebooks and Travel Writing Studies

In popular conceptions of 'travel writing', guidebooks tend to rank highly. Those who are not regular consumers of narrative travel books often assume that a 'travel writer' must be someone who works on guidebooks for publishers such as Lonely Planet and Rough Guides. And yet, the scholarship of travel writing – and in particular the anglophone scholarship of anglophone travel texts – has some difficulty in accommodating guidebooks within its own taxonomies.

Jan Borm's well-known definition of the travel book frames it as 'any narrative characterized by a non-fiction dominant that relates (almost always) in the first person a journey or journeys that the reader supposes to have taken place in reality while assuming or presupposing that author, narrator and principal character are but one and the same' (2004: 17). Other definitions of travel writing provided by Tim Youngs (2013: 3) and Carl Thompson (2011: 26) use very similar parameters. These definitions clearly privilege the long-form travelogue, the extended first-person narrative account of a journey – that branch of travel writing most likely to be accepted as a literary or aesthetic genre.

Where, then, does this leave guidebooks? Typically, they lack an obvious overarching narrative. They also typically lack any obvious use of the first person – except perhaps in a few 'author recommendation' box texts. And any claim for their 'literary' or 'aesthetic' status would likely be viewed askance by many scholars. The difficulty in accommodating guidebooks within standard scholarly taxonomies of travel writing perhaps compounds a degree of ambiguity as to which discipline they rightly belong to as an object of study. For literary scholars, their 'uncertain status as literature' (Mee 2009: 305) is even more pronounced than that of travelogues; while their superficial resemblance to literary texts (as books rather than more ephemeral periodical media) may deter scholars approaching from the direction of media studies or tourism studies. This may explain why, as Scott Laderman has noted, so 'little scholarship exists on the [guidebook] genre' (2016: 258).[1] This is despite its ubiquity, and despite the fact that historically the guidebook was not clearly delineated from the first-person travelogue.[2]

Modern guidebooks do, in fact, often contain traces of latent narrativity, and encrypted elements of the first person, particularly in the moments when a guidebook breaks briefly out of its customary mode of the perpetual present tense to describe recent conditions at a particular tourist attraction, typically using phrases such as 'when we last visited' (Hannigan 2021). But a readier way to accommodate guidebooks within the scope of travel writing studies may be to emphasize in Jan Borm's definition not the use of narrativity and the first person, but the *non-fiction* status of travel writing. In another scholarly characterization, Mary Baine Campbell argues that 'A travel book is a kind of witness: it is generically aimed at the truth. Neither power nor talent gives a travel writer his or her authority, which comes only and crucially from experience' (1988: 2–3).

Guidebooks seem, at first glance, to fit even more readily within Campbell's schemata than literary travelogues. Talent – presumably of the literary variety – is relegated; only that generic targeting of truth, emerging exclusively from experience, matters. This, in fact, is where the literary travelogue becomes somewhat contentious, for there is a broad acknowledgement within the scholarship of travel writing that a first-person travelogue has to be, as Carl Thompson notes, 'a somewhat fictionalised rendering' of the author's journey (2011: 28). The non-fiction status of the guidebook, however, appears far less troubled. These are informational, utilitarian texts, compendiums of facts. A bad or out-of-date guidebook may be accused of being 'inaccurate', but it is seldom described as 'fictional'. But this may, in part, be due to an assumption – an inaccurate one – about the nature of the 'truth claim' of guidebooks, and the relationship of that claim to experience.

## Guidebook Modes and the Mechanics of Guidebook Production

The aesthetic and informational aspects of travel writing – typified by the literary travelogue on the one hand, and the guidebook on the other – have only fully separated during the last hundred years or so. Even in the early twentieth century, travel books sometimes had a curious hybridity. Their narratives were at times redolent of unique, individual experience, but were nonetheless plainly intended to provide a practical guide to other travellers following the same route, with explicit directions and suggestions for overnight stops and places of refreshment. An excellent example of this is the series of guidebooks about the coastline of Cornwall, produced by A.G. Folliott Stokes in the 1900s. These books describe a linear narrative of a journey on foot around the coastline. But they are narrated not in the first person singular, but in the first person plural:

> Having briefly glanced at some of the chief characteristics of England's Ultima Thule, we will now start on our long walk to Land's End. Keeping to our original intention of following the coast, except when making detours into the moorland, we pass along the Digey, St Ives' most characteristic street, and turning left find ourselves confronted by the Atlantic. (Folliott Stokes 1908: 51)

Informational guidebook and travelogue here exist together in the same text. The author invokes a shared reader-writer gaze far more explicitly than in the singular mode of more recent narrative travel writing. While they are explicitly designed as guides for 'those intending to make a complete circuit of the Cornish coast' (Folliott Stokes 1910: 22), at times the nature of Folliott Stokes's books as narratives of the unique and unrepeatable personal experience of their author is almost entirely unconcealed:

> At the bottom of the valley, not far from the shore and close to the stream, is a cottage ... Thinking to gain a little local information we knock at the door. There is no answer, and we find it locked. On passing the front-room window we peep in through the dusty glass. The room has whitewashed walls. In the centre of the floor is a table; close to it, on the window side, is a chair, pushed slightly to one side; and on the table opposite the chair is a large open Bible. On its right-hand page is a half-eaten pasty. (ibid.: 24)

This text is suffused with movement – movement that we can reasonably suppose has taken place in actuality. The writer *knocks*, *passes*, *peeps*. And yet, the use of the first person plural ('we') and the present tense maintains the guidebook-style voice. It suggests that, should the reader visit this particular place and follow the directions provided by the book, act out the same specific movements ('knock at the door ... On passing the front-room window' etc.), they will be able to repeat the experience for themselves. (It is worth noting here that while the 'we' mode was common in Folliott Stokes's day, contemporary guidebooks are more likely to deploy the second-person conditional: 'if you look to the left, you will see . . .') However, the intense granular detail suggests that this is, in fact, a narrative record of Folliott Stokes's own, unrepeatable experience at one transient moment, already irretrievable at the time of writing. It is highly unlikely that a subsequent traveller using his book as a guide would spot the same open bible and the same half-eaten pasty; the cottage curtains might well be closed, and someone might respond to the knock at the door. As a guidebook, the text here fails in its most basic function – to provide instructions leading to a directly repeatable travel experience. It may, indeed, be considered flagrantly inaccurate. Nonetheless, it is unlikely that any reader using Folliott Stokes's work to guide their own Cornish coastal walk would ever doubt that the author had been to all the places he describes.

Conversely to the Folliott Stokes example, when a nineteenth- or early twentieth-century travel book did *not* aim to serve directly as a guidebook, the practice of the writer often had clear commonalities with that of modern guidebook writers. A nineteenth-century travel writer arriving in, say, the Himalayas typically gathered hard data, population details, information about specific journey times, about what was on sale in the markets, and about what accommodation was available for the visitor, just like a modern guidebook writer, but perhaps less like a modern narrative travel writer, soaking up the atmosphere and recording their own inner journey. Andrew Wilson's 1875 book *The Abode of Snow*, for example, is a travelogue, a first-person, past-tense narrative of a journey in the Indian Himalayas. However, the author clearly recognizes that his book may serve as a source of practical information for subsequent travellers. Of Mussoorie, he writes that 'The "Himalayan Hotel" there is the best hotel I have met with in India; and there are also a club-house and a good subscription reading room and library.' He also provides suggestions for excursions and detailed information about routes to onward destinations, as well as narrating his own specific journey: 'The two great excursions from this place are to the Jumnotri and the Gangotri peaks [and] are well worth making . . . It is of importance to the tourist to bear in mind that, in order to pursue his pleasure in the Himalaya, it is not necessary for him to descend from Masúri' (Wilson 1875: 42–43). Once again, few would doubt that the author has visited the places he describes and has verified the information he provides first-hand.

This underpinning assumption of experience-based authority survived the near-complete generic separation of the narrative travelogue and the practical guidebook over the course of the twentieth century. The guidebook researcher as much as the narrative travel writer was expected to write from a position of experience-based authority – that is, to have been to the places they wrote about. Indeed, as recently as 2008, Lonely Planet explicitly reaffirmed the claim, asserting that its writers would check 'every listing in person, every time, every edition' (Calder 2008). Patently, this was never actually true when it came to guidebooks. An updater working for a publisher such as Lonely

Planet typically spent between one and two months on the ground, covering a very large area, including many small towns and villages with just one or two hotel and restaurant listings. It would have been practically impossible for even the most diligent author to make the lengthy journey – many hours or even days – required to visit all those places in person; though, if they *were* diligent, they would hopefully attempt to check from a distance that the businesses were still operating, and to obtain up-to-date prices. This from-a-distance element of guidebook work has clearly become vastly easier in an age of digital connectivity. Tiny mountain guesthouses, until recently contactable at best via an unreliable phoneline, now have websites and TripAdvisor listings. This, in conjunction with falling revenues and a marked contraction of the travel publishing industry – even pre-COVID-19 – has led to the increasing dominance of the 'desk updater' over the on-the-ground guidebook researcher.

At this point I should make it clear that I worked as a freelance guidebook writer for over a decade during this period of change. My earliest guidebook commissions, in the mid-2000s, involved checking circular hiking routes on the ground for new editions of old books – confirming that the routes were still passable, that the café at the end point was still open and so on. My most recent commission, however, just before the COVID-19 pandemic got underway, involved sitting at my desk checking facts via the internet for a guidebook to a country I'd last visited in person seven years earlier. By that stage, this was probably the dominant mode of guidebook work.

It can be reasonably argued – regrettable as it might be for the guidebook writer and disappointing for the guidebook user – that a desk-based fact-check for a new edition of an old guidebook does not involve any particular mendacity. The writer is merely trying to verify that particular hotels and attractions are still in operation, that buses still run along particular routes and so on, all of which can reasonably be done at a distance (it's worth noting, too, that in the 1870s Andrew Wilson sometimes provided practical information about journey times and lodgings on certain Himalayan routes that he had *not* travelled in person, presumably having gathered the information from local informants). An infamous case in 2008 in which erstwhile guidebook writer Thomas Kohnstamm claimed to have written a guide to Colombia for Lonely Planet without ever leaving San Francisco was later revealed to be a confected controversy when it emerged that Kohnstamm had, in fact, only been commissioned to update the introductory sections on history – a job that the publisher never expected to involve actual travel (Siddique 2008).

But guidebooks do also contain descriptive, impressionistic text, in particular the sections that attempt to convey the 'feel' of a place. Here the guidebook appears to move closer, once again, to the aesthetic or literary narrative travelogue. This is where, in the era of the 'desk updater', the nature of the experience from which the writer's authority emerges deserves interrogation.

## Believing Las Vegas: Travel Writing from Virtual Experience

Peter Hulme argues that 'All travel writing involves an explicit or implicit truth claim: writers claim to have been in the places they describe. If they are subsequently discovered not to have been, their work is discredited' (Hulme and Youngs 2007: 3). An atmospheric guidebook introduction, conjuring up the feel of a city or providing an impressionistic account of a particular attraction, does not typically involve a directly revealed narrative 'I'. But, if we accept Hulme's assessment, an implicit truth claim is surely still in place. How else could the author have produced the text other than by visiting the place in person? The authority, as per Campbell's schemata, discussed above, is surely assumed to emerge from experience. With this in mind, I will share an example:

> It's Vegas, baby, the city where anything goes. A galaxy of lights glittering in the desert darkness, this is a place that rings to the rhythm of a thousand slot machines, a vision of neon, baize, and crystal. Whatever you've heard, Vegas will still blow your mind. And it's not all about gambling: you can lap up the Vegas style without ever wagering a cent, taking in world-class food, world-famous musicians, and world-class shows. Meanwhile, just beyond the city limits you'll find some of America's most dramatic landscapes. But when it comes to that old adage 'what happens in Vegas stays in Vegas', forget it – there's no way you can leave this great city without a few stories to tell.

This is the introduction to a digital guide to Las Vegas that I was tasked to write some years ago for an app-based travel guide provided to hotel guests. It is clearly an attempt to provide a colourful evocation of the place and its atmosphere, but here I need to make a professional admission: I have never been to Las Vegas. The work was completed in an office in the UK, and no budget was provided for on-the-ground research. It seems, then, that the implicit truth claim mentioned by Hulme has been violated here, and the authority-emerging-from-experience as stipulated by Campbell is absent. The digital guide I wrote should, it seems, be discredited. But is that in fact the case? We need to consider how I actually put together my descriptive evocations of Las Vegas.

I began the task by buying several existing guidebooks to Las Vegas, and by watching several travel documentaries about the city – which is also exactly how I typically began my research upon receiving a guidebook commission that did involve actual travel to the destination. These texts clearly informed the introduction quoted above. Here travel writing's high degree of intra-genre referentiality, its intertextuality, its tendency to be 'excessively citationary' (Melman 2002: 117) and its capacity to reiterate tropes and discourses over and over come into play. The well-known 'what happens in Vegas stays in Vegas' line, for example, emerges from the imaginative construction of that city to which most of us have been thoroughly exposed through myriad texts and media even if we have never visited in person. Discussing 'deep mapping' practices, Peter Hulme has suggested that travel writing 'is actually most at home with itself when it reveals how [it has never] really left the armchair' (2009: 147). He is referring here to the heavy research component of deep mapping, but this could, in fact, be extended to all informational or ideas-based travel writing in which the account of a bodily journey may be somewhat secondary to the information – historical, cultural, geographical – about the place described.

The digital guide I wrote to Las Vegas also contains entries covering individual attractions in and around the city. These feature practical information about entry fees, opening hours, shopping and dining facilities and so on – all of which could be readily gathered from online sources. But they also include short, impressionistic introductions, which seek to convey the 'feel' of the attraction. Two examples follow below, from the entries for the Hoover Dam and the Scenic Drive in the Red Rock Canyon area:

> Hanging in a cool curvilinear sweep amidst the broken, sun-scorched settings southeast of Las Vegas, the Hoover Dam is an otherworldly vision. Its 700-foot fan of pale concrete melds into a landscape of cracks and fractures, glowing in the harsh desert light. Both an engineering triumph and an art deco masterpiece, it is a place that will stop you in your tracks at every angle.

> The 13-mile Scenic Drive alone is a Vegas essential. The road here is a thin ribbon of asphalt threading through a landscape like the rubble of some cosmic catastrophe, great banks of rock thrown up at crazy angles, with color schemes ranging from rusty red to yellow cream.

Both entries are strongly suggestive of direct observation on the part of the author. If they are able to describe the Hoover Dam's 'cool curvilinear sweep', and the way it 'melds into the landscape', the natural assumption of any reader will surely be that they have been there and observed it directly. More significantly, both texts contain a strong implicit suggestion of movement. The Hoover Dam is 'a place that will stop you in your tracks at every angle', implying that the author has moved around it to view it from different positions, and in doing so has been frequently stopped in their tracks by the spectacle. On the Scenic Drive, meanwhile, the ribbon-like road 'threads' its way through a landscape in which the colours range and change. Again, the implication is that the author has completed this thirteen-mile journey in person, making observations along the way. Knowing that they have *not* done so, we may be inclined to discredit the text. But a question remains: if we accept that the descriptions – including their implication of movement – are not the product entirely of the imagination, how then were they constructed by an author who had never approached the Hoover Dam or driven through Red Rock Canyon?

In the case of the Hoover Dam, the answer lies with Google Street View. While researching the attraction, I made copious use of this function of Google's online mapping services. The complex road system around the dam, with its many hairpin bends and elevated vantage points, allows for a multi-angle virtual tour of the site, and the opportunity for a full 360-degree view at any point. I moved – virtually – back and forth around the dam, pausing on approach roads, on the Mike O'Callaghan–Pat Tillman Memorial Bridge and on the dam itself. While doing so I made notes about the landscape, and the colours and textures I noticed – just as I would have done if researching the site on the ground. This experience of the site involved only a single sense (sound, smell and the sensation of wind and temperature were all absent). But it was a unique and original experience: it is unlikely that any other Street View user would move around the Hoover Dam exactly as I did, pausing at the exact same spots, and noticing and recording the same specific features of the landscape.

When it came to the Red Rock Canyon Scenic Drive, I also made use of Google Street View at key points. But the more important component was a YouTube video, featuring dashcam footage recorded during a complete circuit of the route, and lasting around thirty minutes. As the video played, I again observed the landscape and made notes – of the rubble-like terrain, the angles of the hillsides, the varying colours (again, smell and physical sensations were absent from the experience, but the video did feature an unedited soundtrack of engine noise, gear changes and bumps). This exposure to the Scenic Drive was arguably a more passive one than that of the Hoover Dam: I could not choose my own route, and, beyond the very rudimentary measure of pausing the video, I could not stop and view the surroundings at will. But the experience retained a degree of unrepeatable uniqueness: it is unlikely that another viewer of the video would focus on the exact same passing crags or cactuses.

I deployed similar research practices throughout the writing of the Las Vegas guide. All of this, arguably, can be categorized as *experience* of the place – and as Mary Baine Campbell argues, it is from experience that a travel writer's authority arises. Furthermore, that experience, as I have argued, was unique and unrepeatable. Each reader of this chapter could take a Google Street View journey along the Las Vegas Strip, for example, and each would notice different authentic particulars – the cracks in the tarmac, the crash barriers, the slightly shabby quality of the concrete, the way the leaves of the date palms droop – and then emphasize them differently in their written descriptions. Crucially – and this is the key point – the experience of Las Vegas here is mediated by the technology used to record and transmit it, but it is *not* mediated by the 'seeing eye/I' of any previous traveller, as would be the case while reading a travel book (either a travelogue or a guidebook). In that sense, the experience of Las Vegas is original, and in effect, unmediated. (One might argue that the recorded data is quasi-mediated, in that someone drove the original Google surveying vehicle down the strip; but the automated 360-degree camera mounted atop it did *not* record the driver's view, and each end user will make a unique and original sequence of left-right, back-forth, up-down pans when using Street View.)

My portrayal of Las Vegas was still generically aimed at the truth, and it still, in some sense, arose from experience. I was using the data generated through that experience to produce what I reasonably believed was an accurate account of the actuality – albeit the actuality of a place I had never visited in person. It is worth briefly reflecting here on Dürer's rhinoceros. I had no more direct contact with the actuality of Las Vegas than he had with the actuality of a rhino. And yet the nature of the data sources for our respective depictions was radically different.

## Guidebook Itineraries: A Narrative Amalgam

I want to end by looking at a particular feature within guidebooks, in which the latent narrative becomes explicit: the 'suggested itineraries' that appear in the introductory sections of many guidebook series. These typically offer suggested routes of varying lengths that a visitor might follow: 'One Day in Lyon', 'Ten Days in France' and so on.

Here is an example, taken from a guidebook to Laos and Cambodia that I worked on – another desk update – several years ago:

> **Day 2**
>
> **Morning** Get an early start for the long trip to Battambang. Make a stop at Banteay Srei *(pXX)* along the way if you have private transport; otherwise go straight back to Siem Reap, grab a quick lunch, then hop on a connecting bus for Battambang.
>
> **Afternoon** Shake out your limbs after the long journey with a stroll through the old heart of Battambang *(pXX)*. Admire the Colonial-era shophouses around Pasar Nat and along the riverfront, then continue south to the night market *(Street 1)* at dusk.
>
> **Evening** Wrap things up with a relaxing drink at the Balcony Bar *(Street 1)*.

What is significant here is the form, which is quite distinct from that typically found elsewhere in a modern guidebook. It has explicit narrativity, but its mode is quite different again from that typically found in a travelogue, and indeed from that found in Folliott Stokes's Cornish books: it is delivered in the second person, and in the imperative. This is partly a manifestation of guidebooks' instructional purpose: more generally, they tell people what to see, where to stay and eat. But it also generates a particularly pronounced simulacrum of the experience of travel for a reader. And yet these itineraries – unlike Folliott Stokes's narrative – are typically *not* records of actual journeys, undertaken by their authors. A guidebook writer actually visiting Battambang would not be strolling around or enjoying relaxing drinks; they would be dashing here and there checking opening times and inspecting hotel rooms.

Preparing one of these itineraries as a writer involves creating an imaginary journey through actual places – places that you may or may not have direct experience of. I had visited the places featured in the itinerary quoted above – albeit several years before I worked on the guidebook, and not in the specific order described. However, there were other suggested itineraries in the same guidebook that included stops in places I had *not* visited in person. Again, my information for those sections came from a mixture of other guidebooks, videos, blogs and copious use of Google Maps and Street View – the same blend of mediated and quasi-unmediated virtual experience that produced the Las Vegas guide. The resultant second-person imperative text is, then, a fictional narrative of a potentially actual journey. The journey as described has not been undertaken, but it possesses a rich potentiality: assuming that the borders and roads are open, and the specific eateries are still in business and still serving the same menu items, a guidebook user could undertake the journey exactly as described on the ground in Laos and Cambodia, alchemizing imaginary into actual, despite the fact that the writer who produced the narrative had not made that specific journey described – at least not in the traditional sense of physical movement through a landscape. 'Suggested itineraries' are, then, a form in which the usual travel writing dynamic – writer, journey, text, reader – is dramatically rearranged, with the actual journey, if it occurs at all, undertaken not by the writer but by the *reader*, and *after* the production and reception of the text.

## Conclusion: Travel Writing within a Virtual World?

The fact that I was able to produce a narrative of a repeatable journey without moving bodily along its route, and the fact that I was able to produce an account of Las Vegas drawing on original observations made during a quasi-unmediated experience of the place despite having never been there in person, raises questions about the nature of travel, about the nature of experience and about what it means to have direct access to data. These issues are not new: I have never visited the Louvre, but I first 'saw' the Mona Lisa long before Google Images existed; likewise rhinoceroses. But these questions are becoming rapidly more significant as technology continues to develop. To watch a GoPro recording of a mountaineer's ascent to a summit is to experience that climb vicariously, which might also be true for the reader of a written account; but the GoPro video also, arguably, provides a partially *unmediated* experience – albeit limited in its sensory components to visual and aural. This makes it markedly different from the experience of reading the same mountaineer's written account of the climb. As virtual reality and other digital technologies advance, and as the so-called 'metaverse' develops, the limits of what counts as experience of travel will surely become increasingly unclear.

Charles Forsdick calls travel writing 'the most physical of all literary genres' (2016: 68), in reference to its assumedly inviolable connection with bodily movement on the part of the author. But contemporary guidebook work draws attention to the idea that this direct physicality may not be – and may never have been – quite so central to all travel writing as is sometimes assumed.

Similarly, when Angharad Saunders, writing of the relationship between writing and place, describes a text as both 'a real and imagined entity that is made within the world' (2018: xxi), it is worth considering the potential nuances of 'within the world', and the increasing likelihood that that world might be a virtual one. At the very least, asking of guidebook writers, 'Did they even go there?' may soon be irrelevant, if indeed we have not already passed that point, just as we have with the question of whether it is really necessary to have seen a rhinoceros to produce a drawing of one firmly aimed at the verifiable actuality.

**Tim Hannigan** teaches writing and literature at the Atlantic Technological University Sligo, Ireland. He received his doctorate – based on a critical-creative investigation of contemporary travel writing – from the University of Leicester. He is also an experienced professional travel writer, and the author of several narrative history books with travel themes. His most recent book is *The Granite Kingdom: A Cornish Journey* (Head of Zeus, 2023). His research has been published in the *Journal of Commonwealth Literature*, *Journeys*, *Terrae Incognitae* and *Studies in Travel Writing*, and in various edited collections.

## Notes

1. A recent useful addition on this front is Alasdair Pettinger's survey of guidebooks in *The Routledge Research Companion to Travel Writing* (2019).
2. It is worth noting here that francophone travel writing studies has paid a certain amount of attention to guidebooks, for example with the substantial 2000 collection

*Les Guides imprimés du XVIe au XXe siècle: Villes, paysages, voyages*, edited by Gilles Chabaud, Évelyne Cohen, Natacha Coquery and Jérôme Penez, and based on an earlier conference at Université Paris-Diderot. It is also worth noting that francophone travel writing scholarship seems typically to have fewer anxieties about the 'literary' status of travel texts than its English-language counterpart.

## References


Adams, Percy G. 1962. *Travelers and Travel Liars*. Berkeley: University of California Press.

Borm, Jan. 2004. 'Defining Travel: On the Travel Book, Travel Writing and Terminology', in Glenn Hooper and Tim Youngs (eds), *Perspectives on Travel Writing*. Aldershot: Ashgate, pp. 13–26.

Calder, Simon. 2008. 'Travel Guide Fraud? No, Just Flawed', *Independent*, 19 April. Retrieved 5 July 2024 from https://www.independent.co.uk/travel/news-and-advice/travel-guide-fraud-no-just-flawed-811623.html.

Campbell, Mary B. 1988. *The Witness and the Other World: Exotic European Travel Writing, 400–1600*. Ithaca, NY: Cornell University Press.

Chabaud, Gilles, Évelyne Cohen, Natacha Coquery and Jérôme Penez (eds). 2000. *Les Guides imprimés du XVIe au XXe siècle: Villes, paysages, voyages*. Paris: Belin.

Folliott Stokes, A.G. 1908. *From St Ives to Land's End*. London: Greening & Co.

———. 1910. *From Devon to St Ives*. London: Greening & Co.

Forsdick, Charles. 2016. 'Travel and the Body: Corporeality, Speed and Technology', in Carl Thompson (ed.), *The Routledge Companion to Travel Writing*. Abingdon: Routledge, pp. 68–77.

Hannigan, Tim. 2021. 'Beaten Tracks: Belatedness and Anti-Tourism in Guidebooks', in Maria Månsson et al., *The Routledge Companion to Media and Tourism*. Abingdon: Routledge, pp. 215–24.

Hulme, Peter. 2009. 'Deep Maps: Travelling on the Spot', in Julia Kuehn and Paul Smethurst (eds), *Travel Writing, Form, and Empire: The Poetics and Politics of Mobility*. Abingdon: Routledge, pp. 132–47.

Hulme, Peter, and Tim Youngs. 2007. *Talking about Travel Writing: A Conversation between Peter Hulme and Tim Youngs*. Leicester: English Association Issues in English.

Laderman, Scott. 2016. 'Guidebooks', in Carl Thompson (ed.), *The Routledge Companion to Travel Writing*. Abingdon: Routledge, pp. 258–68.

Mee, Catherine. 2009. 'Journalism and Travel Writing: From Grands Reporters to Global Tourism', *Studies in Travel Writing* 13(4): 305–15.

Melman, Billie. 2002. 'The Middle East/Arabia: "The Cradle of Islam"', in Peter Hulme and Tim Youngs (eds), *The Cambridge Companion to Travel Writing*. Cambridge: Cambridge University Press, pp. 105–21.

Pettinger, Alasdair. 2019. 'Guidance and Advice', in Alasdair Pettinger and Tim Youngs (eds), *The Routledge Research Companion to Travel Writing*. Abingdon: Routledge, pp. 139–52.

Saunders, Angharad. 2018. *Place and the Scene of Literary Practice*. Abingdon: Routledge.

Siddique, Haroon. 2008. 'Lonely Planet Writer Plays Down "Fake" Reviews', *Guardian*, 14 April. Retrieved 5 July 2024 from https://www.theguardian.com/uk/2008/apr/14/10.

Thompson, Carl. 2011. *Travel Writing*. Abingdon: Routledge.

Vertue, Kim. 2020. 'A Rhinoceros-Shaped Idea', *Medium*, 6 December. Retrieved 5 July 2024 from https://medium.com/signifier/a-rhinoceros-shaped-idea-d18d4ca5458.
Wilson, Andrew. 1875. *The Abode of Snow*. London: Blackwood and Sons.
Youngs, Tim. 2013. *The Cambridge Introduction to Travel Writing*. Cambridge: Cambridge University Press.

CHAPTER 10

# Dialogue Journaling in Travel Writing Projects

Clarisse Chicot-Feindouno, Charles Mansfield and Mark Stothard

## Introduction: The Context of the Research

This chapter documents a writing residency in an urban regeneration area in Plymouth, UK. The creative team of three practitioners working on this project had common aims in developing a research practice for literary travel writers that draws on the use of sketchbooking by artists and photographers. Chicot-Feindouno and Mansfield had taught writing workshops in tourism and place-making, from which Chicot-Feindouno had produced a form of metafiction travel writing as her practice. Stothard and Mansfield had encountered the artist's sketchbooking process through postgraduate classes on a master's programme in photography with Dr Carole Baker; Baker's teaching and writing engages with post-humanist moves in creativity (Henning and Baker 2021). The notion that the sketchbook is a process, that it encourages experimentation, allows stages to be worked through with care and, at the same time, records the steps as a genealogy, connected with the project's main aim. This aim was to test a method for field writing projects called dialogue journaling, which will be explained in detail in this chapter for other writers to use.

Working on a discrete project together would give each of the three team members an opportunity to exchange ways of working and to move from pure academic research on writing to a practice that was closer to literary writing but still rigorous and substantiated by data collection. The team comprised an artist-photographer and two writers; the plan was to bring sketchbooking from photography into affective writing to create and test a processual journaling practice. A further aim was to reach an audience that would be wider than the academic community alone; research impact for the public in the UK will again be a major consideration in the Research Excellence Framework in 2029 (REF 2024). Each of the team members was seeking a way of exploring urban space and the shifting Self through experiments with text, inspired by Baker's classes on post-humanism and the following proposal from Deleuze and Guattari on the writing process: 'To write is perhaps to bring this assemblage of the unconscious to the light of day, to select the whispering voices, to gather the tribes and secret idioms from which I extract something I call my Self (*Moi*)' (Deleuze and Guattari 1987: 84).

Thus, the project was both personal and professional. In personal development it was to be a marker placed in the ground that the team members had experienced a period of creative residency. The team had completed funded research projects for universities, but the two writers, Chicot-Feindouno and Mansfield, had not been commissioned for a creative art project before. The individual members needed this experience to be able to approach commissioning organizations for writing residencies in the future with a proper understanding of what text could offer to city projects. Even in this last thought, a residue of tourism development remains, the notion that humans can remake a space via the symbolic realm of literature into a place of pleasure and enjoyment for others.

André Gide opens his short novel *Paludes* (1895) with this Latin epigraph: 'Dic cur hic' ('Say why you are here'). Writing the answer to this, or rather journaling the reason to say why you are here, is the moment when a 'toureme' (Mansfield 2015) begins to emerge. The toureme is the moment when a literary text becomes activated by the visitor's cultural capital when walking in the tourist destination where the novel is set. As early as 1957, in his essay 'The Writer on Holiday', Roland Barthes had remarked upon this pleasurable process of the imagination that occurs when a text places famous writers in a real setting. Barthes himself was not immune to the toureme moment of being in the same place as a famous writer. Documenting this in his autobiography, he declares that he once saw Nobel laureate André Gide (1869–1951) eat a pear in the Lutétia brasserie in Paris (Barthes 1957).

This chapter explores how a methodology of dialogue journaling when working on the move, in the field as a literary travel writer, can catalyse toureme moments and at the same time assist researcher-writers in creating the catalyst for later readers. A full explanation of dialogue journaling is provided in this chapter for other researchers and writers to reuse. The chapter presents processual writing practices that synthesize rich literary affect in the imaginary of the writer. Drawing on Spinoza's *Ethics* (Spinoza 1996), it signals the move from affect to affection for a place. It announces the diffraction of modernist tropes, through deep mapping, to a post-humanist anarchive (Lessard 2009), for the contemporary literary writer. Through Bakhtinian dialogue between the three researchers on the writing residency in Millbay and with tourism stakeholders, a more nuanced and polyvocal text can emerge than through traditional place-making or travelogue (Bakhtin 2010).

Annie Ernaux's two travel books set in Paris – *Journal du dehors* (1993) and *La vie extérieure* (2000) – provide examples of 'small stories' (Patron

**Figure 10.1.** Pedestrianized entrance to Millbay Boulevard. Plymouth, UK, showing the two residency writers and the photographer reflected in the water fountain, 2023. © Mark Stothard.

2020), in which Ernaux uses the commuter train, the RER, as her field and her literary laboratory. One example of Ernaux's presentation of a travel moment as a 'small story', in the way that Patron theorizes this(Patron 2020), occurs in *La vie extérieure*. The narrator, who is Ernaux journaling her observations on the move, recounts a scene, frozen from the perpetual displacement of everyday commuting. In this scene a young couple are waiting for a train that one of them must catch, leaving the other on the platform: 'Le train que l'un des deux devait prendre allait arriver, comme la fin du monde' (Ernaux 2000: 107).

> 13 avril
> Les stations du RER étaient désertes ce matin d'un lundi de Pâques. À Neuville-Université, sur le quai en direction de Paris, seulement un couple qui se serrait en silence, sans un mouvement. De la rame en direction de Cergy, on ne voyait que le dos de la fille. Lui avait la tête enfouie dans son cou. Quand le RER a redémarré, on a pu voir son visage à elle, une fille avec des lunettes. Elle regardait devant elle, loin. Le train que l'un des deux devait prendre allait arriver, comme la fin du monde. (Ibid.)

> 13 April
> The RER stations were deserted this Easter Monday morning. At Neuville-Université, on the platform for Paris, there was only a couple hugging in silence, without a movement. From the train to Cergy, all you could see was the girl's back. He had his head buried in her neck. When the RER started moving again, you could see her face, a girl with glasses. She was looking ahead, far away. The train that one of them was supposed to take was about to arrive, like the end of the world. (Translation by the authors)

The journaling entry from 13 April is developed from a simple observation of a couple on the platform into a narrative event. It has an evocative scene-setting, with the empty train stations and the stillness of the couple heightening the focus and mood. Then the movement of the observer's train reveals the face of one of the figures to provide the readers with characterization. Finally, Ernaux adds a literary simile, 'like the end of the world', to complete the conversion of the scene into a 'small story' as Patron (2020) theorizes this form. Without this literary creativity the scene would have been a documentary text and would not elicit affect. Thus, examples of small stories are incorporated into the new proposed process as plateaus of written moments whereby the everyday documentation of the urban space is frozen in a moment of literary affect (Deleuze and Guattari 1987).

Millbay is an area of marshes and dockland located in Plymouth, UK, overlooking the ferry terminal for France. The project documented here, called the Millbay Residency, worked with Gide's metafiction, *Paludes* (1895), and Damion Searls's 2021 English translation of Gide's original, titled *Marshlands*. The aim was to generate toureme moments that create connections with a literary text for future visitors to enrich their experiences of the regenerated urban

**Figure 10.2.** Refurbished nineteenth-century buildings on Union Street, Plymouth, UK, being prepared for community and social enterprises during the Millbay Residency, 2023. © Mark Stothard.

area. As a tourism aim for the project we envisaged visitors exploring the new urban area on foot, staying at one of the hotels in Millbay and taking the ferry to and from France from the nearby terminal. We therefore wanted to find comfortable walking routes that led to places of hospitality and sites of historic interest and tourist activities.

## The Millbay Residency

Until their merger in 1914, Stonehouse was a separate town, a neighbour to Plymouth in the UK. Since the turn of the millennium, former UK Ministry of Defence property in Stonehouse has passed into council and then private hands for redevelopment of the dockside, including the area known as Millbay. Millbay, as an urban area, is bordered by Mill Bay, a quayside stretch of water on the English Channel where ferries with a tonnage of 42,000 and draughts of up to seven metres can safely dock. The first notable recent leisure project in Millbay was to convert the disused Royal William Victualling Yard into an area of restaurants and tourism activities. Former military and industrial sites in expanding European cities in the early twenty-first century often undergo rebuilding projects to boost their appearance and economic activity. A documented example of this change of use of urban space to provide leisure facilities is Nantes, in western France (Fernández Agueda 2014). Plymouth was experiencing similar investment at the time of the project. Our first mission was to find walking routes into and around this emerging tourist area as it changes use to leisure and becomes part of the experience economy in the 2020s. As the project took form under the title of 'The Millbay Residency', the team of three concentrated their field research on a marshland regeneration scheme funded by the European Union's Interreg initiative, which was reaching completion on Bath Street, Plymouth. Our data collection methods included writing field-notes at stopping points along the streets we explored, taking photographs as both location reference points and as art photography for use in exhibition or publication, and dialogue between the three team members while out in the field or with locals, visitors and workers encountered along the way. Writing on the move was an important consideration for the research. The objects of the research were the experiences afforded by the urban space during the visits, and how these experiences could be recorded and communicated by writing. Coupled with this was the investigation of how writing itself can be an inquiry, leading the researcher not simply to record but to create new narrative knowledge during writing.

The work used dialogue journaling, an elaborated note-taking and note-making process, proposed at the time of the residency as a methodology for urban inquiry (Mansfield and Potočnik Topler 2023). The objective of dialogue journaling is to mobilize Bakhtinian dialogism for the creation of affective texts that can become cultural artefacts in their own right. In Bakhtin's dialogism 'prosaic creativity generally proceeds slowly, begins in narrow spheres, and is hardly noticeable. For that reason, we do not see it, and think that innovation must come from somewhere else. But innovation is in fact the product of innumerable small changes taking place "incessantly." The difficulty we have in perceiving and understanding it results from its very familiarity' (Morson and Emerson 1990: 23). This slow and careful innovation is incorporated into a three-step journaling process, presented in table form below (Table 10.1), which the team adopted for the residency project. This process is reproducible and so generates research data that can be verified by other researchers, and is a methodology that can be applied in other urban place-writing projects. Dialogue journaling moves pure research into research and development, because a cultural object is made during the final synthesis step.

Journaling in this way concerns itself with whatever is happening at the time of the exploration of a space. These different types of knowledge can be as disparate as the encounters of the writers as they move around the city space, the books appearing at the same time as the journaling or articles mentioning the space, the new streets and their emerging architecture, and the flora and fauna coming to colonize these new boulevards and squares. Other writing practices to interrogate the city are already well documented; a collection of essays by Kristin Ross (2023) appeared during the residency, which demonstrates the power of the detective novel to explore urban space. Maigret *polars* and the investigations of Hercule Poirot are practices that were considered as possible writing methods for communicating the experience of being in the city. Ross makes a statement that resonates with the aims of this inquiry in her discussion of detective fiction: 'The space of the city and its various markers each allude to moments in a complex and differential history, to the varying timescales of staggered evolutions of institutions, ways of being, and material space' (ibid.: 224).

The material space of Millbay manifests its complex history as a travel interchange, with the waning fortunes of its previous leisure industry marked

**Table 10.1.** The three steps of dialogue journaling.


| | | |
|---|---|---|
| 1 | Library | Building a literature review and deep-map of the urban space from literary and historical texts. Planning and sketching walking routes from maps. Gathering media assets and content for later blog and social media posts. |
| 2 | Field | Experiencing and recording the environment. Engaging with locals and visitors through questions and dialogue. Building trust with tourism stakeholders through social media channels and blog posting live from the field. This creates a dialogue through the comments added to the social media posts. |
| 3 | Lab | Applying care to the collected texts from earlier journaling. Recounting the story and articulating the findings through synthesis. |

by disused theatres, a closed-down cinema and an empty ice-rink. A short piece of detective fiction, 'The Plymouth Express' (Christie 1923), initially seemed to speak to the chosen space since it was set on a moving train and first published by *The Sketch* on 4 April 1923, when Millbay station would have been the destination for its characters. Christie's short story, though, did not suit this urban exploration because Millbay does not specifically feature in the text. The choice of a literary text to accompany the project was postponed until initial library journaling had been completed to better understand the area and identify an appropriate literary period and genre. Library journaling is Step 1 in the dialogue journaling research process. This step is similar to a literature review in traditional research projects, but the notes taken from referenced material are discussed through written dialogue with other team members, and this dialogue is retained for reference during the design of any walking routes.

## Writing Catalysts

There follow three examples of catalysts taken from dialogue with the place under study or from the chosen text to exemplify how the writing can be set in motion: (i) Rain Gardens and Tree-Planting, (ii) First Passengers to Plymouth and (iii) Say Why You Are Here.

### *Rain Gardens and Tree-Planting*

A European urban space was inaugurated in Plymouth in June 2021: Millbay Boulevard. New pine and silver birch trees were planted, and an underground flood defence system began to regulate the flow of rainwater into the sea in Mill Bay during the research period; and on Tuesday 13 June 2023 the first new business opened, the two-hundred-room Moxy Hotel. The boulevard protects the surrounding community from flooding by the buffering capacity of tanks below the rain gardens, which are visible in the pavements. A safe, dry urban space emerged from the marsh ground around Bath Street. New buildings appeared slowly, contrasting with the workshops that had once served the old Plymouth Millbay railway station. The station was used for goods trains running through to the docks until 30 June 1971. Millbay was beginning to come back to life as a space of travel and tourism.

The bay itself is in the tidal estuary of the River Tamar, the border with Cornwall. This stretch of the Tamar even has its own name, Hamoaze, an Old English toponym that means water meadow in the mud. During the residency, and so an event that could be incorporated into the journaling, a story of the border created by the Tamar and the travel writer's decision of whether to take the leap and cross it was published (Hannigan 2023). The research team decided not to make the crossing from Devon despite the temptation of the Cremyll Ferry, which would have taken us to a little piece of Cornwall where another writer was just publishing his story of journaling (Gale 2023) while our own writing-on-the-move project unfolded.

### First Passengers to Plymouth

If you stand at the front door of the old Duke of Cornwall Hotel, which opened in 1865, and look north across Millbay Road, that was where the entrance to the passenger terminal of the railway station stood. In 1936, when trains could already reach speeds of 100 miles per hour, Great Western advertised the station and the ocean dock with the slogan 'Land at Plymouth and save a day'. Transatlantic passengers disembarking at Plymouth from liners bound for London via Le Havre, Normandy could save this time by taking the train from here, a journey of only three hours fifty-four minutes. The last transatlantic passengers stepped from ship to train in 1963. After a decade without international passenger traffic, on 2 January 1973 – the day after the UK joined the European Community – Brittany Ferries began their service from Roscoff. Then, on 24 January 1974, Brittany Ferries brought their first passengers to Plymouth on board their new vessel, *Penn-Ar-Bed*. However, by then Plymouth's railway station was well over a mile away, an uphill walk of thirty-three minutes.

**Figure 10.3.** The New Palace Theatre of Varieties, Union Street, Plymouth, UK, rebuilt in 1899 after a fire, closed in 2006. Grade II listed building, classified at risk, 2023. © Mark Stothard.

### *Say Why You Are Here*

As a creative team of travel writers and a photographer our aim was to create or find a reason to say why we had chosen to write on the move in this space. We wanted a pretext to offer new visitors the potential for a toureme to emerge (Mansfield 2015). Thus, the writing was for and about a conditional future. The hope was to create a type of literary text that visitors would enjoy and take on board as part of their own personal cultural capital. This is referred to as making a text that is a cultural artefact in its own right, in the same way that a novel is a cultural artefact. During the first or library step of the three-step process researchers seek out a novel or literary text set in the destination. In this case none could be found. Yet a series of signs were beginning to point towards the work of André Gide (1869–1951). Gide had been Barthes's inspiration as the concept of the toureme began to germinate. One of the creatives in the team had worked in metafiction, which is one of Gide's literary forms, and the damp marshlands of the Millbay area echoed the setting of Gide's sparkling travel story, *Paludes*, now available in a recent English translation as *Marshlands* (Gide 2021). This timeliness of an English version being available so close in time to the residency research gave it extra value for journaling because the journaling process aims to capture key moments in the present that will become historical for later readers. One of the affects of a toureme moment is a type of nostalgia, but a nostalgia of not being there at the time and imagining an era when all those past events created a situation that is recorded and offered up by the text being read.

A deeper connection with Gide's writing process was also at work in the residency. One of the aims was to experiment with how to write on the move, and to this end we were employing the process of dialogue journaling (Mansfield and Potočnik Topler 2023); full details of dialogue journaling are given in the following section, and in Table 10.1 above. One of the aims was to pilot dialogue journaling using the free platform offered by Google Docs, Google Drive, Google Keep and Chat Space, and to use free technologies to transcribe both handwritten text and recordings of any interviews and dialogue between the researchers. Gide is studied for his journaling process. Indeed, his novel *The Counterfeiters* (1925) is paralleled by his *Journal of The Counterfeiters* (1927), which for writing researchers gives an insight into his thoughts and life during the production of the novel, and for travel writers gives licence to record the everyday creative process during their writing projects. In *Paludes*, Gide lets himself be inspired by Virgil's *Eclogues*, originally written 44–38 BCE (Virgil 1916), and cleverly inserts extracts – the word eclogue can mean extract – of a novel that the narrator of the book is writing as the travel story of *Paludes* unfolds. This internal, yet explicit, discussion of the writing process places Gide's *Paludes* in the genre of metafiction. Virgil's *Eclogues* contain a dialogue on land use between Meliboeus and Tityrus and are relevant for our writing project on Millbay; Virgil's characters even enter into dialogue with themselves. Here, for example, Meliboeus tells himself what he must do: 'Meliboeus: Now, Meliboeus, graft your pears, now set / Your vines in order!' (Virgil 1916: 1.73). The conversations we had with each other and the notes we took from these while on the move would accumulate to try to answer the question that Gide poses at the opening of his novel *Marshlands*, 'Dic cur hic' (Gide 1895: n.p.), which translates as 'Say why here', rather than studying any other place.

## Dialogue Journaling with Purpose

The first purpose of dialogue journaling is to free the travel writer for easy note-taking in the field followed by more considered note-*making* back at the writer's desk, which we refer to as the lab in this process. Taking notes in the field, in this method of journaling, tries to capture as many observations, feelings or spoken comments as the researcher can feasibly manage to note down or record. Including such thorough data collection in the methodology was prompted by Paul Rabinow (2007). In his ethnographic research in Morocco, Rabinow only realizes much later the importance of the 'swirl of people, places and feelings' (ibid.: 6). Rabinow illustrates this with Hegel's metaphor, 'the owl of Minerva flies at Dusk' (ibid.: 7), which means that the researcher does not know which data from the field will ultimately reveal meaningful themes when interpreted later. Thus, the research team aimed to take notes to create a large amount of data, hoping that the themes would take flight later in the lab; here you can see the ethnographic genealogy of our approach, as currently practised in qualitative research methodologies in social science disciplines: 'Researchers dive deep into the content to discern initial themes and important sections. They then select quotes that bring the data to life and aptly represent diverse viewpoints' (Naeem et al. 2023: 2).

In Gide's *Marshlands*, the short novel that was chosen as a catalyst literary text, the narrator describes the same concern for gathering data from the moments of travel. On the morning of their journey the narrator asks, in self-dialogue: 'Now, what should I say about that moment? Why not talk about it at least as much as about the following moment – do we know, after all, which things are the important ones? What arrogance there is in choice! Let us regard everything with the same insistence' (Gide 2021: 73).

As Paul Ricœur (2007: 25) explains, in hermeneutics we do not know what we will later interpret from events and where we will find meaning. On the move, we therefore take notes as often and as thoroughly as possible. To that end, the team equipped themselves with hardback, lie-flat, spiralbound notebooks from Hamelin in Normandy so that they could write even when standing and when no hard surface was available on which to rest. Pencils or rollerball pens were used to avoid rain smudges. For dialogue journaling it is important to date and time entries in the journaling field notes, with a note of the weather at the time, too. During the fieldwork the researchers learnt to write in more print-style handwriting so that the smartphone scanner, Google Lens, could transcribe fieldnotes more accurately. If lines were too narrowly spaced, then the software would mix text from adjacent lines. The project therefore used Hamelin's Oxford A5 ruled notebook rather than the company's French-style squared preprint. Further, if a word or phrase is crossed out – placing it in the category of Derrida's *sous rature*, 'under erasure' (Derrida 1976) – then Google Lens finds it and transcribes it anyway, so we try to remember to add square brackets around a crossed-out phrase.

## The Three Steps of Dialogue Journaling

The methods for writing during this urban residency were very deliberate and self-conscious since the process of dialogue journaling was being tested to establish it as a methodology that other researchers and practitioners could use. The first step in dialogue journaling for literary travel writing is a period spent in the library. In traditional research terms this echoes the literature review. One of the expectations for writing affective literary text, though, is to find existing affective texts that are linked to the urban space of the destination under study. A critical and emotional reading of the discovered texts is recorded in the first of the three folders of the journaling notebook. For this project, an earlier template for a Microsoft OneNote notebook was adapted into a series of Google Docs documents. Each document was considered as a notebook started by one of the residency creatives but was made write-accessible to the rest of the team for contributions to the section designated as Dialogue. The research process examined a range of texts and documents in this step, most particularly the material from the European Union Interreg project describing the building of the urban drainage scheme that was named Millbay Boulevard. A further source was the press release material from the hotel chain, the builders of the new infrastructure and the ferry company that docks in Mill Bay. Geological and geomorphological information on the site was retrieved from British Geological Survey resources. A more literary or affective mode of travel writing, from the point of view of the aims of this work, should express the changes in land use in the places it explores. This is close to the concept of deep-mapping, where the travel writer or place-making writer uncovers the human history of a site, along with the effects of the climate, botany and geology of the place. The term vertical travel writing (Forsdick 2020) also conveys this idea of the researching writer digging deeper down into the layers of time and geology and upwards into the atmosphere to present the place as more than a simple, flat, two-dimensional place.

Step 2 in dialogue journaling is the fieldwork itself. Typically, a travel journalist will only have one or two days in the field to complete a writing assignment. Tour guides, too, when designing an urban walk, will consider the collection point for their group of visitors – the hotel or transport hub – then the main visitor attraction at the furthest point of the guided walk, with an elliptical or circular route back – stopping for appropriate comfort breaks – to the hotel or coach. This period of residency research was rather different, in that the researchers returned many times to the field either in the team of three, or alone, carrying with them copies of the catalyst text, *Paludes/Marshlands*, and later an English translation of Virgil's *Eclogues* (1916). Below is an example of how these catalyst texts stimulated the journaling. Less urgency provided time for discussion as we walked and explored different routes to, and then through, the space. The movement through the streets of Millbay began to draw and delimit what became a mapped place for the researcher, in much the same way that W.G. Sebald emotively maps out a delimited space of mobility in Vienna in his classic literary travel text, *Vertigo*:

> Early every morning I would set out and walk without aim or purpose through the streets of the inner city, through the Leopoldstadt and the Josefstadt. Later, when I looked at the map, I saw to my astonishment that none of my journeys had taken me beyond a precisely defined sickle- or crescent-shaped area, the outermost points of which were the Venediger Au by the Praterstern and the great hospital precincts of the Alsergrund. If the paths I had followed had been inked in, it would have seemed as though a man had kept trying out new tracks and connections over and over, only to be thwarted each time by the limitations of his reason, imagination or will-power, and obliged to turn back again. (Sebald 2002: 33–34)

In trying to find parallels with the urban exploration at work here, the most connected antecedent is the practice of Michèle Bernstein, one of the original Paris Situationists from the mid-1950s. For Bernstein the work was not completed with a *dérive*, which is a self-conscious walk across a city. Bernstein wanted to go further, by creating literary artefacts from the psychological experience of moving through the streets, which she achieved with the novel *Tous les chevaux du roi* (1960) [*All the King's Horses*]. The novel itself is also a *détournement*, a process appropriated by the Situationists, to rework existing cultural artefacts to serve new ends. Bernstein has Gilles, one of her characters, express this in the novel:

– De quoi t'occupes-tu au juste ?
– De la réification, répondit Gilles. . . .
– Je vois . . . C'est un travail très sérieux, avec de gros livres et beaucoup de papiers sur une grande table.
– Non, . . . je me promène. Principalement, je me promène. (Bernstein 1960: 16)

– 'What are you doing exactly?'
– 'Reification', Gilles replied.
– 'I see . . . It's a very serious task with huge books and lots of papers on a large table.'
– 'No, . . . I walk. Mainly, I walk around.' (Translation by the authors)

Of course, soon after her *dérives* through the streets and galleries of Paris, Bernstein then wrote literary works, principally this *roman à clef*, a genre close to metafiction where the characters and situations can be inferred as if from a key to real people and places. In the current research process developed here, the aim was not only to gather experiences of the city in the same way that the much-discussed figure of the *flâneur* might do, but also to pass on to a step of synthesis and creation, as did Bernstein by the end of the 1950s. To give the final text a literariness, during the lab step in the process, again Bernstein's creative synthesis provides a precedent in that she uses past tenses in the original French to lend veracity to the situations presented. Occasionally she even moves into the *passé simple* tense, which is difficult to simulate in English, but indicates strongly yet subtly that a story is being recounted rather than travel advice being imparted.

In our own practice, as soon as possible, each researcher would scan and add the fieldnotes to the formatted pages online in a shared Google Drive folder. Here is an example, inspired simultaneously by Gide's use of Virgil's *Eclogues* and Gide's exhortation to write poetically when in the field – a discovered extract from Virgil that has links with the new hotel accommodation that opened on Millbay Boulevard during the period of the residency:

TITYRUS:
Yet here, this night, you might repose with me,
On green leaves pillowed: apples ripe have I,
Soft chestnuts, and of curdled milk enow.
And, see, the farm-roof chimneys smoke afar,
And from the hills the shadows lengthening fall! (Virgil 1916: Eclogue I)

Virgil's Latin verse, rendered here as pentameters in the English translation, generated this response as a dialogue from the team's writing on the move:

> Step 2 Field Journaling Entry: 12:40, Tuesday 27th June 2023.
> Boulevard
> Here water, channelled and stored, moves beneath the boulevard to assemble itself as a rhizome by finding lines of flight in new pine trees that grow from its underground body.

Later, on 19 July 2023, a more careful piece of journaling was developed from this very short initial field entry: 'I stood for a long time on the front steps of the Duke of Cornwall Hotel looking across to where Millbay Station stood. The last train ran through here on 30 June 1971' (Millbay Field Journaling 2023: 6). Spending time later that day, when sitting in better working conditions, and with library access, this was developed thus. Notice how 'the front steps' is changed to 'the top step' to initiate the sense of the narrator looking down on the scene:

> I stood for a long time on the top step of the Duke of Cornwall Hotel looking down to where Millbay Station once stood. The last train ran through Millbay on 30 June 1971 on its way to the dockside where I had seen the preserved tracks on Soap Street, earlier on that morning's walk. The hoardings that protected the works around the new Moxy hotel had been taken down for its official opening on Tuesday 13 June 2023. This broad, sweeping vista revealed a direct walking route from here above the old station through a narrow set of steps back onto Millbay Boulevard. It was a new ope, the Plymouth word for an alley. What would it be called? Station Ope? Pear-Tree Ope, Moxie? Moxie, spelt with an -ie ending, is a word from the Abenaki language meaning 'dark water'. The Abenaki first-nation peoples would have encountered the settlers from the Mayflower when it completed its journey from Plymouth in November 1620. (Millbay Residency Journaling 2023: 6)

In this longer extract above, we have also shifted into the past tenses in English. Again, this is a deliberate practice to create the storytelling effect, and to present the I-narrator as having been in that location for the readers. In the original French of *Paludes*, Gide also uses the *passé simple* tense which indicates to readers that a story is being recounted. We strove to reproduce that story-like feeling for our English readers.

One method created for gathering new stories of place we have called Recounting Steps. Bartlett's experiments with ghost memory from the 1920s and 1930s, as explained in Wagoner et al. (2019: 16–18), furnish a ready-made method for retelling experiences from memory by case-participants, or in our case by the writer-researchers themselves. Wagoner et al. show how the reteller makes qualitative changes as the experience is retold. It suggests that human memory is not a storage archive, like a computer, but is transformative, in that it integrates the reteller's own culture as it is mediated appropriately for their listeners. The reteller is thus an active agent, which, for this research, can be considered an enrichment of the recalled experience. In this research the re-counting stage was repeated after the experience to allow this enrichment to develop.

The whole team of three sit in a café in the urban space under investigation. Two are chosen to walk to two new locations found on the map, less than 400 metres distant. They have a set time, usually twenty minutes, to explore the new place then return. On their return, the team sit with voice transcription software apps running and question those who have been on the reconnaissance trip. The aim in this micro-method is to elicit the reconstruction of space into words by each informant. In the face-to-face situation of demand the informants process their experiences into past tenses and into a form that the listening interlocutor can understand. In fact, the interlocutor, as the research revealed, asks further questions in dialogue to try to better understand what is being reported.

The recounting method evolved from an earlier concept called the writing plateau. The writing plateau is conceptualized twice in Deleuze and Guattari's *A Thousand Plateaus*. Brian Massumi in his translator's foreword explains his own use of the term: '[A plateau in writing is when the] heightening of energies is sustained long enough to leave a kind of after-image of its dynamism that can be reactivated or injected into other activities, creating a fabric of intensive states between which any number of connecting routes could exist' (Massumi 1987: xiv).

Deleuze and Guattari themselves consider their contributions to the collective work of compiling the book as operating as plateaus. And within the realm of this research a plateau is a flat stopping-point in the fabric of the city, a place of ease for the researching writer but also one that offers a point of experience or a situation that can be captured by journaling. Plymouth has many slopes since it has grown up along riverbanks and beside the sea, so finding a flat area of terraced ground to stand comfortably to talk and to write requires the discovery of plateaus. Often the writing plateau is indoors in a public space such as a café, hotel reception area, gallery or shop. In the same way that tour guides find safe, flat spots on which to stand to address their audience, the researcher-writer in our process does the same. From the vantage point of the plateau, lines of where the next step can be taken open up before the researcher. Thus, these pauses to complete some journaling on a plateau give the researcher time to carefully consider the next direction to take and, as such, allow the urban space to enter into dialogue with the researcher. For example, an interesting sign or building might draw the researcher's attention to walk towards it. The plateau, though, will retain its after-image in the journaling that took place there, and may find its way into the final writing to act as a toureme for the readers of the final piece.

Step 2 in the field also provided us with opportunities for trust-building with the range of cafés, hotels, galleries, workshops and the local authority. For this the researchers used social media. In the field one of the team used the free platform Google Maps, which offers write-access via a social channel called Google Local Guides. Through Google Local Guides reviews were written and posted in real-time of the cafés used as plateau places for writing and talking with the staff. We found that we could better explain what we were doing when we could show the public-facing staff our review posts to this social media channel. Many small businesses on our routes already had a presence on Google Maps and so appreciated the reviews we posted on the move. This allowed us to build relationships for return visits and further dialogue on what this emerging area meant for their lives.

To build trust with larger, more remote stakeholders, longer posts were published in magazine-style blogs. Although carefully constructed back in the lab rather than written live in the city, these posts were also timely and released when members of the team were walking in the field again. The example below, from the photographer's blog, shows the attempt to mobilize interest in the project using evocative travelogue-style language. The timeless present tense is deployed alongside imagistic description to create affect in the readers:

> In the heart of Plymouth, tucked away amidst the historical tapestry of Devon's coastline, lies the enigmatic precinct of Millbay. Here, within the cradle of time, each cobblestone and aged edifice whispers tales of a bygone era, painting a canvas that resonates with the essence of centuries past. As a lens wielder capturing moments suspended in time, Millbay stands as a testament to the convergence of history and modernity, a muse that beckons with its rich narrative. (Stothard 2023: 1)

Blog posts like the above provide reference points, specific web pages with static URLs that researchers could post to specific social media channels to engage a different demographic group. For example, the URL for this post was posted to LinkedIn to make connections with the public relations departments of the hotel chains in the area, the ferry companies and the local destination marketing organization (DMO). As magazine travel writers know, these press office staff can provide researchers with a rich resource of historical information about their companies' operations in an area, but trust must first be established. The LinkedIn post with the URL for this blog text gained seven hundred impressions, over 27 per cent of them from employees of large companies.

In addition to LinkedIn, the residency work also used Facebook, Instagram, Mastodon, Threads and, as previously mentioned, Google Local Guides. This post, below, to Local Guides for example, was to locate the project by journaling a precise date to say that the researchers were here by using the past tense:

> Café Bené on West Hoe Road is clean, stylish and serves good food promptly. Mark and I called in here for brunch on Tuesday 7th March 2023 during our fieldwork to explore the EU-funded Millbay Boulevard project. This is the urban regeneration drainage and tree-planting scheme to protect this area of Plymouth from flooding. The café is well-positioned near Fin Steps to connect to the water's edge in Soap St. Search for Fin Steps on Google Maps. (Guide 2023: 1)

At the time of writing, this post with photographs to Local Guides has had 2,800 views and the post for Cawfee Café on Union Street has received 322,000 views. When compared with the views of the team's own blog posts, which average about sixty views in the year, the Google Local Guides posts provide much greater exposure for urban writing projects while they are still live.

Step 3 takes place back in the lab. This step is the opportunity to synthesize a new literary artefact, a story set in Millbay. Using examples from the residency, this section shows some of the ways the literary writing is produced, taking care to draw out themes from the catalyst text so that the final story of Millbay has a connection with both the place and the chosen novel. In this way the new Millbay story contains potential touremes for readers to discover for themselves.

In the following extract from a more developed text, the opportunity to enrich the Millbay travel story comes from a greater familiarity with the chosen catalyst text, *Paludes/Marshlands*. In dialogue meetings, the team agreed to include *Marshlands* and some of its themes hidden beneath the story. The conceit was that the marshes of Millbay are now hidden beneath the new boulevard and new hotel, so the created text would be a metaphor for this regenerated urban space. Here is an extract of the created Millbay story as a work-in-progress during Step 3 in the writing lab. The narrator is waiting for the ferry from Plymouth to Roscoff, France:

> I held my breath for a while, I had no need to become engaged with this place of transition, I thought. All I had to do was get through the waiting. I stood for a long time on the top step of the Duke of Cornwall hotel. My view swept down through the old Millbay station to the water course that is now beneath the boulevard. Melt water, silt and pebbles had rushed down this track for eleven thousand years; on the move to the Atlantic as it filled to form a channel to separate Brittany from Devon. The edge of the ocean inched its way up to the water course. The mill wheel turned more slowly. The ocean waited.

> In the evening at the hotel, Gide would be tempted outdoors to walk to the Palace Theatre of Varieties, in Union Street. Was Angèle with him? When travelling through a strange town you can afford to risk entertainment that you wouldn't try at home. Pleasure while travelling, is only an accessory to the education it brings. (Mansfield, Millbay Lab Journaling 2024: 2)

Gide's text, *Marshlands*, and later his 'Billets à Angèle' (Gide 1921), deploy the character of Angèle as an interlocutor for the narrator, probably as a reference to Virgil's characters in dialogue, Tityrus and Meliboeus. For the final Millbay story, we wanted the project to include pointers to the chosen novel by Gide and some of his practices. Gide develops a subtle relationship between the two main figures in *Marshlands*, where the character of the narrator is not fully aware of Angèle's intelligence and often over-explains his ideas to her; the final sentence in the above extract is an echo of one of those explanations, which readers of *Marshlands* who are visiting Millbay could notice and enjoy the connection.

The above longer section gives exemplifications for each of the three steps of the whole process to help readers see how they might adopt this for their writing projects. It is hoped that this section demonstrates that during place-making work or residencies for creatives, taking time to engage in dialogue with locals and with published texts linked even tentatively with the place can enrich the archive. With a rich digital archive, the final Step 3, or lab step, is more likely to help the writer produce stories with greater affect for the potential visitor.

## Writing in Motion

In *Marshlands* the narrator is writing, persuading others to travel, and then travelling. These three activities were pursued in the Millbay Residency. Using literary travel writing, the work of the project was to explore what could be done with writing in motion. Gide's narrator is writing the novel-within-a-novel to persuade his fellow writers to travel in order to change their routine lives. Tityrus in the meta-novel keeps finding reasons to be content with his routine life and not travelling to another location. Overall, though, Gide is writing for pleasure, his own creative pleasure, and to share an ironic yet meaningful cultural artefact with his contemporaries, one that they would enjoy.

Our experimentations in writing about place had drawn from the work of Deleuze and Guattari (1987). When approaching a new urban space, the team initially aimed for a route design of six plateaus. We aimed to find and choose six flat places set on an elliptical route around and through an urban space to provide the researcher with stopping points where it was warm, dry and flat enough to write. Many months into the Millbay Residency we realized that discrete visits to the field of inquiry, often separated by several weeks, resulted in one or more researchers suddenly appearing at new midpoints. One morning, for example, one of our team (Mansfield) arrived by car, rather than walking down from the new station, and could not find a parking space behind the community shop called The Plot on Union Street, so he set foot in our field of inquiry upon a remote and unknown plateau. This must be the experience for anyone travelling through Millbay; the visitor suddenly emerges from a taxi or a car, or even from the Roscoff ferry, and, although they are at their starting plateau, they are already in the middle of things, *in medias res*. As Deleuze and Guattari note, 'A plateau is always in the middle, not at the beginning or at the end. A rhizome is made of plateaus' (ibid.: 21). In his introduction to *A Thousand Plateaus*, Massumi also draws attention to this concept of sudden emergence: 'Nomad space is "smooth," or open-ended. One can rise up at any point and move to any other. Its mode of distribution is the *nomos*: arraying oneself in an open space' (Massumi 1987: xiii). Here is the journaling stored in our Google Docs system of that moment at a new open plateau:

> Like Kafka in his journaling, I was able to develop an idea about everyday work, and thus answer Gide's question 'Why have I witnessed this scene?'
>
> Why did I come in here? Why here? Aristotle thought that plot was built on reason. A text must tell its readers why, for it to have plot, or at least the literary text must pose the question why, leaving that mystery in its readers' minds.
>
> My intended reasons for exploring the café area of the newly opened hotel were to see if writers could meet and work in the environment. Would it be easy to find a table that was not too low for writing? Would it be too noisy? Would there be hidden pressure to drink up and go? Then, as I sat to sip my delicious cappuccino, I noticed that the new hotel's glass curtain walls turned the Devon morning sunshine into a silvery white light.
>
> Bathed in this cool light, and way across the huge seating area someone opened her laptop and began typing. Papers surrounding her made me conclude that she had been working undisturbed for some time. I was going to be able to write in this new space that had only existed for 5 weeks.
>
> Outside, up on the car park, I knew that two young birch trees and a pine were settling into their new home, too. The silver birch (*betula pendula*) is a pioneer species, reclaiming difficult urban wastelands to

> support insects and plants that shelter beneath its canopy. The silver birch is a nurse tree creating a space of care around it for others to flourish. Betulin, the white compound in the birch's peeling bark, protects the tree from any strong winter sunshine, reflecting the radiation and sharing the light with those nearby. Silver birch grows mutually with the mycorrhizal fungus, *amanita muscaria* to connect its roots with an underground supply network, tree, and rhizome in symbiosis. I realised that the working laptop user must have a wifi connection. (Mansfield, Millbay Lab Journaling 2024: 1)

Another academic taking a line of fleeing from conventional research writing, Christopher Hanley, figures with the concept of double articulation from Deleuze and Guattari to find an experimental method of writing: 'The aspects of written assemblage I want to emphasise are internal organisation and co-functioning. As we saw above, the double articulation comprises the form and substance of content and form and substance of expression' (Hanley 2019: 416). Is Hanley proposing a new way of writing for pleasure, as our research suggests that Gide is doing in *Marshlands*? And, of course, one of the aims is to create a type of text that generates experience of Millbay, Plymouth: 'I will be approaching the text as a generative space, in the sense that an engine generates movement. I will be theorising how words and signs give access to the data, prompted by discussion of Deleuze and Guattari's concept, "regimes of signs"' (ibid.: 417).

This coincides with the aim of making the output text, *Millbay Boulevard*, into a report of the research process even though this final output should be a pleasure for visitors to carry with them and read when in Millbay. Functioning like Gide's idea of a pretext, the final piece has other pre- and post-signifying texts circulating around it; the aim is to create a text that can accommodate these different regimes that surround it. The extracts must be open to the pleasure that this interplay will generate in our readers. This is very close to Hanley's figuring with generative text and will set the writing in motion when it is enjoyed by readers as visitors in the field. The concept that writing is a movement itself, most particularly the process of journaling or note-taking in the field, is expressed in the other academic work that appeared during the time of the Millbay Residency. Below, Ken Gale is discussing his diaries, his own journaling notebooks in this work: 'So, they often appear, come to life and live in action amongst the messiness of a cluttered bar-room side or pushing aside the coffee cups on a café table when a thought comes to mind, when the urgency of an idea is too intense to resist and when the immanence of the movement/moment is perpetuated by writing in expression' (Gale 2023: 148).

## Conclusion

Very positive benefits emerged from using Web 2.0 digital notebooks during this project. The first was the realization, just a few days into the journaling, of the freedom that it gave members of the cooperating team to capture anything and everything while working, especially in the field. During the work the team realized that they had collected a wealth of minute detail that was dated, timed and had geographic locations and thus supported later writing up of more publishable texts. The captured texts of the journaling can also furnish the data required for a more traditional qualitative study of the urban space using the methods proposed recently in Naeem et al. (2023). This release from constraint that is realized when researchers can scan handwritten text and combine it in a digital notebook, which is searchable, is sufficiently productive for us to propose that the dialogue journaling method is valuable for travel writing practitioners and researchers.

The initial constraints were overcome through learning how the three-step structure of the model provides a discipline of storage and activity type. Once this is incorporated into the writer's practice then the separate sections become helpful because they provide ready-made places for composing notes into the three steps of library, field and lab, as shown in Table 10.1 above.

It became apparent that the writing team could spend time on the finished text by copying journaling from the fieldnotes section across to the lab section so that original ideas were not lost and time could be spent in a more comfortable writing environment later. This release from urgency creates a feeling of care towards the composition of the final pieces alongside care towards the self. As other travel writers emulate this approach, they will enjoy the process because it allows time for the emotional and affective responses they might want to share with their future readers. These affective expressions are not only of sentiments but also of sensibilities. Sensibilities emerge when writers have time to appreciate the intellectual themes and metaphors presented by their chosen catalysts, catalysts that may be other literary texts, music associated with the place, local botany and local food cultures, and finally through rereading the archived discussions in the dialogue boxes.

The formatted template of each journaling page, too, provides spaces that encourage inquiry, dialogue with others in the project, and later data mining for emerging themes. The template is shown below with annotations on how to use it:

By adopting a dialogue journaling approach to writing projects, especially projects that have periods of being on the move and engaging with wider groups, the writer-researcher will discover exciting benefits of increased confidence, more tangible trust-building with an audience and finally a compelling

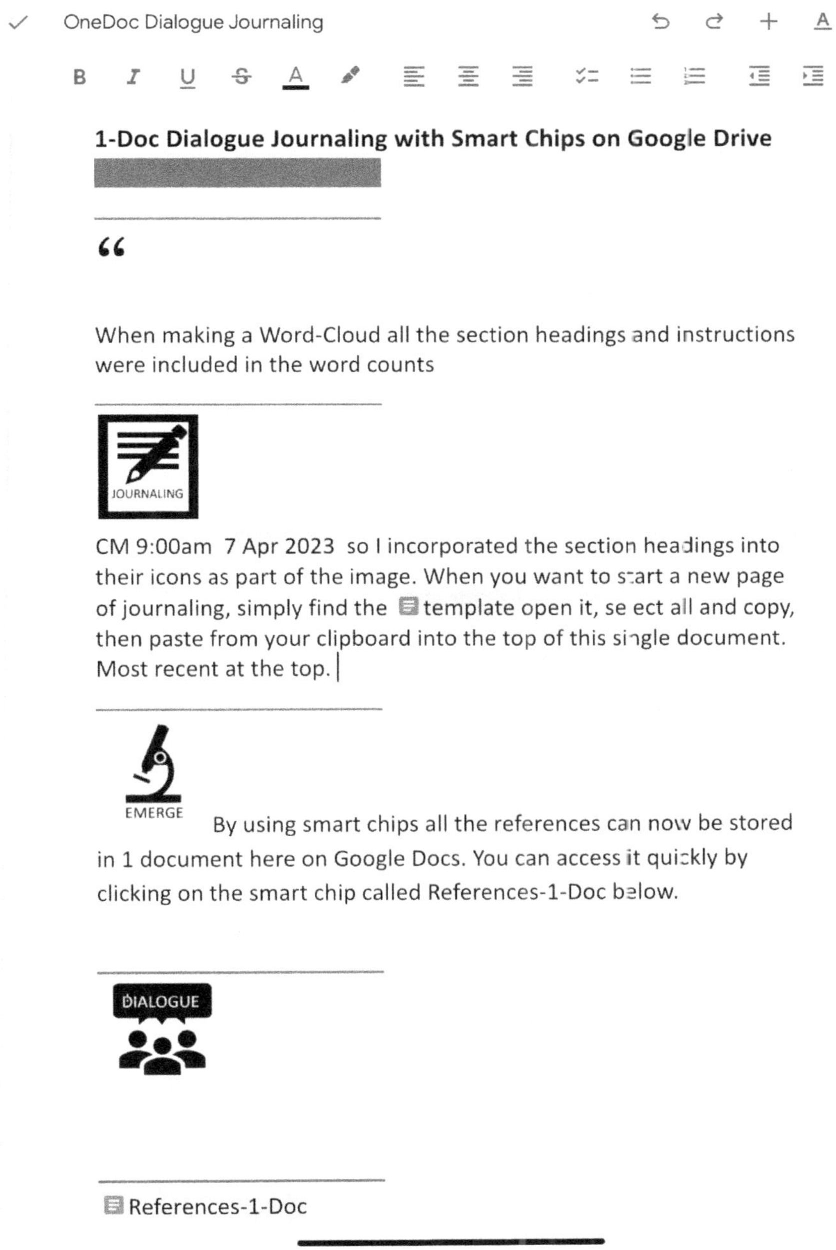

OneDoc Dialogue Journaling

**1-Doc Dialogue Journaling with Smart Chips on Google Drive**

"

When making a Word-Cloud all the section headings and instructions were included in the word counts

JOURNALING

CM 9:00am 7 Apr 2023 so I incorporated the section headings into their icons as part of the image. When you want to start a new page of journaling, simply find the template open it, select all and copy, then paste from your clipboard into the top of this single document. Most recent at the top.

EMERGE

By using smart chips all the references can now be stored in 1 document here on Google Docs. You can access it quickly by clicking on the smart chip called References-1-Doc below.

DIALOGUE

References-1-Doc

**Figure 10.4.** Layout of the page template for dialogue journaling further developed during the Millbay Residency, 2023. © Charlie Mansfield

sketchbook to share, which is an accessible archive of their process and discoveries of the places explored.

**Clarisse Chicot-Feindouno** is an emerging literary travel writer, having studied tourism management at the University of Plymouth, UK. She has worked in the genre of metafiction and explores how this form can be adapted to travel literature and blogging.

**Charlie Mansfield** is an academic and author specializing in methodologies for place-making and literary tourism. He recently completed an EU project on ethnobotany in tourism web-logging on the Adriatic coast. He has previously completed a CNRS-funded project on heritage management and big data for the University of Paris 1, Panthéon-Sorbonne. In summer 2024, he was commissioned under the UNESCO Cities of Literature initiative to write his Millbay experiences as a work of metafiction for translation by the British Council's Connections through Culture project in Indonesia.

**Mark Stothard** is a professional and research photographer. He successfully completed his MA in Photography during the Millbay Residency and is currently working on a major project to develop narrative photography methods in the Scottish Highlands. He is an educator and runs workshops, most recently on Lundy Island, off the coast of North Devon.

## References

Bakhtin, Mikhail. 2010. *The Dialogic Imagination: Four Essays*. Austin: University of Texas Press.

Barthes, Roland. 1957. *Mythologies*. Paris: Éditions du Seuil.

Bernstein, Michèle. 1960. *Tous les chevaux du roi*. Paris: Buchet-Chastel.

Christie, Agatha. 1923. 'The Plymouth Express: A Hercule Poirot Short Story', *The Sketch* 1575, 4 April.

Deleuze, Gilles, and Félix Guattari. 1987. *A Thousand Plateaus: Capitalism and Schizophrenia*, trans. B. Massumi. Minneapolis: University of Minnesota Press.

Derrida, Jaques. 1976. *Of Grammatology*, trans. G. Spivak. Baltimore: Johns Hopkins University Press.

Ernaux, Annie. 1993. *Journal du dehors*. Paris: Gallimard.

———. 2000. *La Vie extérieure*. Paris: Gallimard.

Fernández Agueda, Beatriz. 2014. 'Urban Restructuring in Former Industrial Cities: Urban Planning Strategies', *Territoire en mouvement: Revue de géographie et aménagement* 23–24: 3–14.

Forsdick, Charles. 2020. 'Vertical Travel', in Alasdair Pettinger and Tim Youngs (eds), *The Routledge Research Companion to Travel Writing*. Abingdon: Routledge, pp. 99–112.

Gale, Ken. 2023. *Writing and Immanence: Concept Making and the Reorientation of Thought in Pedagogy and Inquiry*. Abingdon: Routledge.

Gide, André. 1895. *Paludes*. Paris: Librairie de l'art indépendant.

———. 1921. 'Billets à Angèle', *La Nouvelle revue française* 16: 462–66.

———. 2021. *Marshlands*, trans. D. Searls. New York: New York Review Books.

Guide. 2023. 'Café Bené on West Hoe Road', *Google Local Guides*. Retrieved 18 April 2025 from https://maps.app.goo.gl/wBqKDzNwWWKuSBfu7.

Hanley, Christopher. 2019. 'Thinking with Deleuze and Guattari: An Exploration of Writing as Assemblage', *Educational Philosophy and Theory* 51(4): 413–23.

Hannigan, Tim. 2023. *The Granite Kingdom: A Cornish Journey*. London: Head of Zeus.

Henning, Michelle, and Carole Baker. 2021. 'The Subject as Object: Photography and the Human Body', in Liz Wells (ed.), *Photography: A Critical Introduction*, 6th edn. Abingdon: Routledge, pp. 211–63.

Lessard, Bruno. 2009. 'Between Creation and Preservation: The Anarchive Project', *Convergence* 15(3): 315–31.

Mansfield, Charlie. 2015. *Researching Literary Tourism*. Plymouth: Shadows Media.

Mansfield, Charlie, and Jasna Potočnik Topler. 2023. *Travel Writing for Tourism and City Branding Urban Place Writing Methodologies*. Abingdon: Routledge.

Massumi, Brian. 1987. 'Translator's Foreword: Pleasures of Philosophy', in Gilles Deleuze and Félix Guattari, *A Thousand Plateaus: Capitalism and Schizophrenia*. Minneapolis: University of Minnesota Press, pp. ix–xv.

Morson, Gary Saul, and Caryl Emerson. 1990. *Mikhail Bakhtin: Creation of a Prosaics*. Redwood City, CA: Stanford University Press.

Naeem, Muhammad et al. 2023. 'A Step-by-Step Process of Thematic Analysis to Develop a Conceptual Model in Qualitative Research', *International Journal of Qualitative Methods* 22. https://doi.org/10.1177/16094069231205789.

Patron, Sylvie. 2020. *Small Stories: Un nouveau paradigme pour les recherches sur le récit*. Paris: Hermann.

Rabinow, Paul. 2007 (1977). *Reflections on Fieldwork in Morocco*. Berkeley: University of California Press.

REF. 2024. *The Research Excellence Framework in the UK*. Bristol: Research England. Retrieved 14 August 2025 from https://2029.ref.ac.uk.

Ricœur, Paul. 2007. *The Conflict of Interpretations*. Evanston, IL: Northwestern University Press.

Ross, Kristin. 2023. *The Politics and Poetics of Everyday Life*. London: Verso.

Sebald, Winfried Georg. 2002. *Vertigo*. London: Penguin Vintage.

Spinoza, Benedict. 1996. *Ethics*. London: Penguin.

Stothard, Mark. 2023. 'Milbay Project', Mark Stothard MA ARPS personal website. Retrieved 27 June 2025 from https://markstothard.com/millbay-project/.

Virgil, Publius. 1916. *Eclogues, Georgics, Aeneid*, trans. H.R. Fairclough. Cambridge, MA: Harvard University Press. Retrieved 27 June 2025 from <https://www.theoi.com/Text/VirgilEclogues.html#1.

Wagoner, Brady, Ignacio Brescó and Sarah H. Awad. 2019. 'Theories of Constructive Remembering', in Brady Wagoner, Ignacio Brescó and Sarah H. Awad, *Remembering as a Cultural Process*. Cham: Springer, pp. 11–21.

CONCLUSION

# Writing on the Move

## *About and Beyond Travel Writing*

Samia Ounoughi

### Writing on the Move and Meta-Discourse

How do travellers proceed on their journey while writing an account of their travel? To what extent can travel writing be completed on the move? These were the main questions we raised when we launched this project. One way to answer them is the study of meta-discourse, for 'writing on the move' can be construed as 'travel writing in the making'. The metalinguistic aspect of research implied by *Writing on the Move* takes us back to the roots of language, which has a metalinguistic function as language always says something about itself (Jakobson 1980 and 1981). There have been relatively few publications on the meta-discourse of travel writing (Hyland 2018, Diani 2023). Meta-discursive analysis in this volume has led to close observations of the traveller-writer's self-reflexiveness on their travel (Chapters 2, 3 and 8), their writing and their writing conditions (Chapters 2, 7, 9 and 10) and even their motivations for writing (Chapters 2, 3 and 10). Beyond metalinguistics, our contributors have resorted to different approaches to literature (genre studies, gender studies, geopoetics, corpus discourse analysis) and to other disciplines (history, geography, natural language processing) to examine how mobility relates to the act of writing. Their work raises questions of stasis and movement both in travel writing and in travel. They have also analysed the means for travel, including transportation modes and financial means.

Another initial plan for this volume was to present a comprehensive range of corpus types. The contributors have carried out case studies of travel writing (Chapters 2 and 5), compared pairs of travelogues (Chapters 4, 6, 7 and 8) or explored larger travel writing corpora (Chapters 1 and 3). The different corpora equate with a diversity of approaches. Some chapters explore the diachronic evolution of travel and travel writing production modes (Chapters 1 and 9). Others have identified the key role played by an author or co-authors in the emergence of a new subgenre of travel writing as a result of novelty in the travel mode or the travel writing mode (Chapters 2 and 10). Meanwhile, comparing pairs of travel writings and wider corpora shows the rich variety of intertextual and interspatial analyses not only through travel texts but also other documents like notes, scrapbooks, newspaper articles, pictures or moving pictures (Chapters 1, 3 and 8). Moreover, multi-scalar analyses within and across texts has led to close examinations of discursive frames and motifs and shed light on the conditions of travel writing elaboration. *Writing on the Move* also stands out for 'hands-on research' since some of the contributors are also travellers and travel writers (Chapters 1, 9 and 10). All of them have dedicated a part of their research to the study of their own travel writing process.

## The Means of Transport

The travelogues studied in this volume involve different means of transport including walking, climbing, canoeing, driving and taking the train. Analysing them provides information on writing conditions, especially whether the traveller can or cannot materially take notes, draw sketches or take photographs as they proceed on their journey. Some means of transport (canoeing, walking, driving) are operated by the traveller. They require bodily effort and are sometimes dangerous, so that writing en route is not feasible. In such cases, travel semantically merges with its etymological root *travail*, which translates in a travelogue in the pervasiveness of bodily sensorial memories (Chapters 2, 4, 5, 6 and 8). Yet stasis and movement are not always opposed. Writing and displacement can happen simultaneously, if the journey takes place on a vessel in which the traveller can be seated to write (the train, a coach, a ship, a plane) and capture their immediate impression. In any case, investigating the production of travel writing according to the writer's displacement mode is crucial to understand the poetic and structural features that characterize the travel text.

Examining (loco)motion also tells us about the traveller-writer. The choice of a means of transport can partake of the travel writer's self-assertion, for example, women driving a car across continents in the 1930s (Chapter 4), a female traveller sailing down the Nile on a felucca or another refusing to ride a mule up the mountains, going against local traditions (Chapter 6). Analysing how a means of transport specifically relates to the act of writing is a paramount aspect of travel writing studies. It allows us to measure the time and distance between the travel and the travel account. It explains the choice of a specific genre like correspondence, dialogue journaling or sketchbooking, which all have different influences on the structure and poetics of the travelogue. A means of transport is part and parcel of the author's creativity in travel writing, as shown in Stevenson's claim that his *Inland Voyage* brings about a new perspective in river navigation travel writing (Chapter 2). Sensations like the speed of a motor car or the slowness of a walk translate into narrative temporality. Travel narratives of road trips, for instance, will contain passages in which the narrative accelerates or slows down (Chapters 3, 4 and 5). A narrative is a process in both temporal and spatial terms (Ricœur 1983–85).

## Money Issues

Money is another topic that has emerged in this volume. Travelling has a cost, and in order to set off and to move on, travellers need money. Guidebooks provide much information about food prices, landmark and museum admission fees, public transport fares and so on. In general, people who travel need a lot of money. Sadly, they need even more money to pay people smugglers when they travel illegally. Having the means to travel is a core issue because money can make the difference between motion and stasis; it is as simple as that. Yet it seems that money has attracted but little attention in travel writing studies, with a notable – and perhaps unsurprising – exception being a discussion by Alasdair Pettinger (2013) in a contribution to a collection of travel writing studies entitled *Travel and Ethics* (Fowler, Forsdick and Kostova 2013). More broadly, money in literature and art is often a taboo. A number of contributions in this volume draw attention to this loud silence (Chapters 3, 4, 5 and 7).

Travel writing itself can be the resource to afford travel. Szołtysek explores writing on the road as a means for journalist Annemarie Schwarzenbach to make money. As she sent unrevised episodes of her travel feuilleton to be published in the press, her writing was the very means she depended on to keep meeting the expenses of driving to Afghanistan. The *travel/travail* common etymology is here again very significant, for *travail* means both effort and labour, and the French word for work is *travail*.

In order to make money out of their writing, authors also need to gain popularity, cultivating the art of being talked about from abroad, and going places to integrate into close literary circles. Travelling for a writer is therefore a potent career booster, as was the case for Stevenson (Chapter 2). Bouzonviller shows that F. Scott and Zelda Fitzgerald's travels to France are a great example of self-promotion (Chapter 3), and that Paris was the place to be if one wished to be ushered into a network of well-known writers. The portrait of the artist abroad in the American press is also a means of self-promotion through travel for the writer, especially when they appear using modern means of transport like cars. At least, such was the case at that time. Travelling is analysed as a social marker of success and popularity. Even today, social elites are sometimes referred to as jetsetters, which again indicates that using the fastest means of transport to travel the world is an attribute of those perceived as successful. Means of transport are thus revealing of the traveller's profile as a writer. Further travel writing studies in this direction should produce valuable findings.

In literary travel writing studies, little is said about the relative cost of travelling in different destinations, and yet the cheap cost of living in a particular country is definitely a motive to go there. Among the travelogues analysed in this volume, at least four concern destinations where the cost of living is clearly cheaper than in the traveller's European or North American home country. For example, Swiss travellers go to Afghanistan via the Middle East (Maillart's *The Cruel Way*, 1947), while British and Irish travellers spend leisure time in Ethiopia or in Egypt (Murphy's *In Ethiopia with a Mule*, 1968 and Mahoney's *Down the Nile: Alone in a Fisherman's Skiff*, 2007). In this volume, the examples of the cheap destination phenomenon spread across a period of over sixty years, suggesting that this is an enduring motivating feature of travel. It would therefore be worth looking in this direction in future interdisciplinary research in travel writing and mobility studies.

## Mobility and Referentiality: Travel and Literal Immobility

One of the major motives of this book project was to question how mobility relates to travel writing, and to bring out the complexities of an equation that might appear obvious at first sight. The analyses of modes of transport have brought forward the crucial role of motion in travel writing and more broadly in a writer's career. Our contributors have also put mobility and travel into perspective by tackling immobility in travel and travel writing. In order to open new perspectives on the idea of immobility as the opposite of mobility, Hannigan rather suggests the hypernym *stasis* (Chapter 1). Indeed, while immobility primarily refers to a permanent state, 'the quality or condition of being immobile or irremovable . . . incapable of moving or being moved' (WordReference 2025), *stasis* refers to a fixed physical position that does not necessarily last. Another antonym of motion worth analysing is *halt*, for halts are also part and parcel of a travel experience and their functions are diverse (Delaitre 2024). Therefore, studying mobility and stasis or halts in travel writing is a means to transcend the mobility versus immobility dichotomy.

There are many reasons that travellers are led to stop en route because travelling can be a strenuous process and the journey can be interrupted by numerous obstacles. The hiker often feels bodily pain (Chapter 6); an accident may damage the traveller's vessel (Chapters 2 and 7). Travellers can also be hindered at customs offices for not having the necessary documents or for sanitary reasons (Chapter 9; Pépy 2018). Motion can become particularly arduous in extreme weather conditions, so much so that the traveller can come to a standstill, which is often narrated in mountain travel accounts (Chapter 1; Ounoughi 2018). In polar travel writing, Antarctica is literally referred to as 'the Ice' (Chapter 7). Thus, trapped between their departure point and their destination, travellers are often caught in a state of in-betweenness, where standards and norms vanish and give way to renewed personal experiences and relations with the other and the world. Our contributors (Chapter 6; Ounoughi 2018) have therefore resorted to the ethnographic and anthropological concept of liminality (Van Gennep 1969), analysing the poetic and linguistic creativity it spurs.

Conversely, halting partakes of the pleasure of travelling and it is a synonym of contemplation (Chapter 1). Landscapes have notably inspired Romantic poetry and paintings along with many other forms of graphic arts. Beyond contemplation, halts are moments of observation, for instance taking measurements, sketching or mapping (Chapters 1 and 10). Writing on the move, and any form of graphic achievement that complements the act of writing, contributes to shaping the world through one's linguistic formulation of it (Wolff 1997).

Shifting the *mobility versus immobility* dichotomy to *mobility versus stasis* leads to a close semantic investigation into the lexical field of travel, with such words as *journey*, *itinerary* or *route*. For instance, some travelogues have been completed without the author travelling at all. This observation, mainly by Hannigan in this volume, goes a long way back, to the early modern epoch (Chapter 1). It goes even well beyond, for doubts about the veracity of travel and of things seen have been present since antiquity. In 7 BCE, Strabo already stated that 'all travellers are braggarts' (Strabo 1917: 111). There have been as many cabinet travel writers as there have been cabinet geographers. Today, with the powerful tools of the internet, travel writing can also be completed without physical displacement beyond that of the mouse by the desktop (Chapter 9). As Hannigan points out, virtual travel can offer wider command over large areas with videos shot with onboard cameras fitted on drones (Chapter 9). The guidebook author completes a journey with an itinerary but no bodily travel.

This exploration of travel, travel writing and stasis has transcended the initial *mobility versus immobility* dichotomy by highlighting its complexities throughout the evolution of travel, and of space data collection and access, which cannot but impact the poetics of travel writing. Such investigations will no doubt expand in future works of travel writing scholarship.

## Truthfulness and Travel Writing: Referentiality, Fiction and Virtuality

The discussion of referentiality and the travel writing genre is long-running and well-known. Peter Hulme, Tim Youngs and Jan Borm (Hulme and Youngs 2007; Borm 2004) argue that if a journey has not actually been undertaken, then a narrative is fictional and does not belong to the travel writing genre. However, debate on this point continues to the present day (Ounoughi, Peraldo and Quaireau 2025). *Writing on the Move* is a project that intends to transcend the factual versus fictional dichotomy about travel writing in various perspectives.

As a referential type of document, travel writing has been and still is a rich resource of geographical knowledge (Bourdon 2020). Many of the writers in the corpora studied by the contributors to the present book were members of the Royal Geographical Society (Chapter 1) and other academies around the world (Chapters 7, 8 and 10). Travel writing has also very much contributed to expanding our knowledge in many sciences. As a result, truthfulness about the travelled place has been a key value of a travel account. The contributors of this volume have studied data collection in the field and its accuracy, as well as intertextuality, as means to enrich the documentary quality of travel writing (Chapters 4, 7, 8, 9 and 10).

Truthfulness is also a matter of properly recording. Our contributors often note the time gap between the travel and the publication of the travelogue (Chapters 2, 4 and 8). The meta-discourse on making notes, as well as the letters and articles sent out while en route all serve truthfulnes because memory is fallible and immediate recording helps in this respect (Chapters 2, 3, 4, 5, 8 and 10). Such attachment to documenting the referentiality of a journey therefore often implies multimodality as the travel text intertwines with visuals of various natures, such as drawings, paintings or photographs. Travel writing therefore takes the broader sense of both semiotic and semantic graphic accounts of the traveller's experience and impressions. Throughout the centuries the evolution of modes of transport and of graphic production, from drawing and painting to photography and film, complementing a text in print or online, has considerably enriched the production conditions of travelogues. The addition of films to a blog or an online travel guide, for instance, literally presents rather than represents the travel experience through moving pictures (though it is important to note that visuals too can be manipulated). These multimodal resources have broadened the range of supports and arts at stake (Chapters 1, 3, 8, 9 and 10).

Yet truthfulness does not only equate to the referentiality of the travelled place. Informative travel writing about the travelled place has gradually shifted towards the truthfulness of the traveller's experience, and more so in the twentieth and twenty-first centuries (Chapters 1, 2, 4, 6, 7, 8 and 10). This observation is in line with the conception of travel writing as autobiographical. As a result, the same journey undertaken by two travellers ends in two completely different accounts (Chapters 4, 8 and 10). At this stage, travel writing can drift away towards geographical imaginary and reach what Doubrovsky, Lecarme and Lejeune (1993) call autofiction.

A very original perspective is the challenge to the dichotomy between dream and reality in travel writing. This analysis of travel writing reveals that the traveller's movements are imprinted forever and not only transform the traveller when they have completed their travel, but continue to evolve throughout their whole life. As a result, oneiric activity in a travel text reveals the construction of the text and challenges chronology to explore how the traces of the travel experience endure, not only in the narrative but in the traveller's body forever (Chapter 8). A travel experience has consequences deep down and gives the author material to write with the poetics of interspatiality and intertemporality. The travel itinerary is completely reshuffled in the travel text, where events and places coexist in the same passages to tell the outcomes and consequences of the travel experiences as a heritage, parts of which connect with the rest of the traveller's life, beyond the travel experience itself and its immediate aftermath (Chapters 5 and 8). All these approaches put the dichotomy between fictional and referential travel writing into perspective and herald new perspectives in travel writing studies.

## Multimodality and Truth Claims

Travel is often – and in a very literal sense – about sightseeing. As a result, many travelogues contain visuals of various natures, contributing to the effort to convey an impression of the travelled place to readers beyond the written narrative of the account (Chapters 1, 3, 7, 8, 9 and 10). The latter represents the semantic, the linguistic bulk of the travelogue, while visuals represent the semiotic material constituting the travelogue. The semantic and semiotic intertwining and interplay has been a focus of the research here. Hannigan reminds us that the visuals inserted in travelogues have not always been truthful illustrations of the travelled place (Chapter 1). Those who authored the visuals sometimes never saw the travelled place. Yet the quest for accuracy in representation evolved with travellers who sketched and painted what they saw en route. Technical advances like photography were developed in such a way that the equipment became more and more easily portable. As a result, the few who could afford to travel produced more and more multimodal travelogues that became a window open on the world to those who stayed at home.

The digital era is key to multimodality but multimodality existed long before any digital device was ever invented, let alone the internet (Lyons and Ounoughi 2020). Scrapbooks are a very good example (Chapter 3), as are collections (cabinets of curiosity) and maps: all these semiotic traces that nourish travel accounts were there already. Digital devices have however brought major evolutions to travel writing by giving the travel account the density of a presentation more than a representation. Although an account can never replace the travel experience itself, the prefix 're' in *re*presentation has been considerably challenged. Such innovations in travel writing have been achieved through the speed and distance at which travelogues can be diffused to the greatest numbers both privately and publicly on blogs and social media thanks to the internet, broadband and now fibre connection (Chapter 9). Travellers can share their travel as they experience it (Chapter 4). Elsewhere, the stillness of the visuals is now overtaken by posts with moving pictures in which the soundscape is perceptible (Schafer 1977). These are major evolutions.

Digital devices have also rendered travel accounts much more interactive. Reader response, for instance, has always existed, with many published critiques in the press. The digital era has democratized the publication of reader response with the comments that readers can make, not only saying how they liked the account but also adding in their own experience or plans to travel the same place as the writer (Chapter 7). Digital devices, while very useful for sharing a travel account online, have also empowered the travel experience with enhanced perception. For instance, some devices render humans capable of commanding sights much larger than what the eye could see. This is the case with drones. The traveller's attention and their sight are assisted by cameras. A traveller wearing a GoPro camera might close their eyes, but the camera is still recording.

This book contributes to shifting to another dichotomy: that of physical versus virtual travel. It is not because a journey is not actual that it is fictional; virtual travel is an experience.

## Traveller Travels and Texts Travel

Our contributors have shown that travel writing goes far beyond a travel account. The process of travel can also deeply alter the travelled spaces. Such is the case of the seafront of Nice on the French Riviera. British tourists had a major influence in its creation, as they had repeatedly demanded the construction of a panoramic seafront road. It was inaugurated in 1824 and it is no wonder that the French named it the Promenade des Anglais. Bouzonviller shows that in the 1920s, the French Riviera became a place where it was fashionable for American writers and artists to reside (Chapter 3). Therefore, writers on the move contributed to the popularity of the place and influenced the way it came to be perceived in time. This is not only the case for the Riviera but also for the Alps, which were looked upon as dangerous and home to many demons (Bourdon 2011, 2018) before the mid-eighteenth century when Romantic writers and artists began to regard the Alpine landscapes as awe-inspiring (Braham 2004; Debarbieux and Rudaz 2010). Travel writing as a cataphoric projection of a future tourist place has even become a field of action research (Chapter 10).

The stakes of travels and their influence on artists' and writers' productions are scrutinized in light of the scrapbooks and articles that Bouzonviller analyses in Chapter 3. What is important to note is that the elements of the puzzle that shed light on the production of novels such as *Tender is the Night* and *The Great Gatsby* are not necessarily to be found only in F. Scott Fitzgerald's correspondence, but also in his wife Zelda's scrapbooks. This means that their travel experience as a couple has left biographical traces common to both Zelda and Scott. It is very original therefore to notice that while a travel experience is generally considered to be unique to the traveller, in this particular case, the scrapbooks written by Zelda provide enlightening information on the genesis of Scott's work. Again, this raises questions about major perspectives in the definition of travel writing as autobiographical (Lejeune 1975). Similarly, the stereoscopic (Forsdick 2009) and *auto*biographical travel writings of Maillart and Schwarzenbach (Chapter 4) each also contain biographical material on the other travelling author.

## Writing on the Move beyond Travel Writing

When we launched the 'writing on the move' project, we received proposals from fellow researchers who did not work on travel writing but still felt that our call for papers resonated with key aspects of narratives in other literary genres, both referential and fictional. Elizabeth Bouzonviller is a specialist of F. Scott Fitzgerald and her cross-check analysis of the paratexts (Scott's correspondence, Zelda's sketchbooks) demonstrates that classic works of literature that are not typically categorized as travel writing may nonetheless be deeply rooted in their authors' actual travels. F. Scott Fitzgerald wrote *The Great Gatsby* during his stay in France. In this respect, this collection has opened new perspectives, showing that the concept of 'writing on the move' extends well beyond travel writing. It would be very interesting to analyse in a broader project the footprints and traces of actual authorial travels in the shaping of other literature and art. In order to explore what Genette (1987) calls the thresholds of the text, we hope that more work will be dedicated to the analysis of travel-related paratextual material. This will bring forward the major states and influences of intertextuality proper to the author on their own literary or artistic productions.

Some travels are meant to be written on the move but not necessarily as travel literature. In recent decades, writing on the move has been democratized by the digital: anyone – with literacy and an internet connection – can write their travel account on a blog (Cardell and Douglas 2016) or on social media. They can do this publicly or privately. The travelled distances can be very far but they can also be very close (Lyons and Ounoughi 2020). Where, then, is the literarity of travel writing? Have literarity and literality become antinomic? Hall (Chapter 7) emphasizes the scientific stakes of Antarctica, with researchers across a wide range of disciplines producing texts based on observations on this continent intended to broaden our knowledge in disciplines such as geology, zoology or astronomy. These are not travel accounts; they belong to scientific literature, and yet they are also writings on the move (Chapter 7). Beyond this volume, 'writing on the move' is a concept that both includes and helps to conceptualize (most) travel writing, but could also potentially include almost any other kind of text (based on its conditions of production). We hope that these reflections will lead studies in travel writing to bring together more interdisciplinary research across genres.

**Samia Ounoughi** is a senior lecturer in English linguistics at Université Grenoble Alpes. She is a member of LIDILEM (Linguistique et Didactique des Langues Étrangères et Maternelles). She is also a member of LABEX ITTEM (Laboratoire d'Excellence Innovations et Transitions et Territoriales en Montagne) where she works with geographers, cartographers and historians. Her research deals with the relations between language and space, and she specializes in corpus discourse analysis of mountain travel writing. She co-edited 'Exceptions and Exceptionality in Travel Writing' (a special issue of *Studies in Travel Writing*, 2020) with Anne-Florence Quaireau. Her other publications tackle the process of naming features of mountain relief ('Referential Conventions as Compromise', in *Reference: From Conventions to Pragmatics*, John Benjamins, 2023). She is the co-editor of *Twenty-First-Century Perspectives in British Travel Writing: Decentring Epistemologies* (Routledge, 2025) with Emmanuelle Peraldo and Anne-Florence Quaireau.

## References


Borm, Jan. 2004. 'Defining Travel: On the Travel Book, Travel Writing and Terminology', in Glenn Hooper and Tim Youngs (eds), *Perspectives on Travel Writing*. Aldershot: Ashgate, pp. 13–26.

Bourdon, Étienne. 2011. 'Vision biblique et lecture mystique de l'espace alpin', in *Le voyage et la découverte des Alpes: Histoire de la construction d'un savoir, 1492–1713*. Paris: Presses de l'Université Paris-Sorbonne (PUPS), pp. 486–98.

———. 2018. 'Montagne et liminalité dans la culture judéo-chrétienne, xve–xviie siècle', in Marie-Christine Fourny and Stéphane Gal (eds), *Montagne et liminalité: Les manifestations alpines de l'entre-deux XVIe–XXIe siècles*. Grenoble: PUG, pp. 25–44.

———. 2020. 'Comment penser les savoirs géographiques à l'époque moderne (XVe–XIXe siècle)?', *Revue de géographie historique* 17–18. https://doi.org/10.4000/geohist.286.

Braham, Trevor. 2004. *When the Alps Cast Their Spell: Mountaineers of the Alpine Golden Age*. Glasgow: In Pinn.

Cardell, Kylie, and Kate Douglas. 2016. 'Travel Blogs', in Carl Thompson (ed.), *The Routledge Companion to Travel Writing*. Abingdon: Routledge, pp. 298–307.

Delaitre, Carole. 2024. 'No Going Back: Interrupted Journeys and Identity Crisis in Marie Ndiaye', in Charles Forsdick et al. (eds), *Microtravel: Confinement, Deceleration, Microspection*. London: Anthem Press, pp. 65–76.

Diani, Giuliana. 2023. 'Metadiscourse and Non-Academic Genres', in Begoña Bellés-Fortuño, Lucía Bellés-Calvera and Ana-Isabel Martínez-Hernández (eds), *New Trends on Metadiscourse*. London: Palgrave Macmillan, pp. 223–44. https://doi.org/10.1007/978-3-031-36690-1_9.

Debarbieux, Bernard, and Gilles Rudaz. 2010. *Les Faiseurs de Montagne: Imaginaires Politiques et Territorialités; XVIIIe–XXIe Siècle*. Paris: CNRS Editions.

Doubrovsky, Serge, Jacques Lecarme and Philippe Lejeune (eds). 1993. *Autofictions et Cie*. Paris: Université de Paris X.

Forsdick, Charles. 2009. 'Peter Fleming and Ella Maillart in China: Travel Writing as Stereoscopic and Polygraphic Form', *Studies in Travel Writing* 13(4): 293–303.

Fowler, Corinne, Charles Forsdick and Ludmilla Kostova (eds). 2013. *Travel and Ethics*. Abingdon: Routledge.

Genette, Gérard. 1987. *Seuils*. Paris: Editions du Seuil.

Hulme, Peter, and Tim Youngs. 2007. *Talking about Travel Writing: A Conversation between Peter Hulme and Tim Youngs*. Leicester: University of Leicester.

Hyland, Ken. 2018. *Metadiscourse: Exploring Interaction in Writing*. London: Bloomsbury.

Jakobson, Roman. 1981. 'Poetry of Grammar and Grammar of Poetry', in *Volume III: Poetry of Grammar and Grammar of Poetry*. Berlin: De Gruyter Mouton, pp. 87–97. https://doi.org/10.1515/9783110802122.87.

Jakobson, Roman. 1980. 'Poetry of Grammar and Grammar of Poetry: (Excerpts)', in *Poetics Today* 2(1a): 83–85. https://doi.org/10.2307/1772353.

Lejeune, Philippe. 1975. *Le Pacte autobiographique*. Paris: Seuil.

Lyons, Agnieszka, and Samia Ounoughi. 2020. 'Towards a Transhistorical Approach to Analysing Discourse about and in Motion', in Caroline Tagg and Mel Evans (eds), *Message and Medium: English Language Practices across Old and New Media*. Berlin: De Gruyter Mouton, pp. 89–111. https://doi.org/10.1515/9783110670837-008.

Mahoney, Rosemary. 2007. *Down the Nile: Alone in a Fisherman's Skiff*. New York: Back Bay Books.

Maillart, Ella. 2013 (1947). *The Cruel Way: Switzerland to Afghanistan in a Ford, 1939*. Chicago: Chicago University Press.

Murphy, Dervla. 1994 (1968). *In Ethiopia with a Mule*. London: Flamingo.

Ounoughi, Samia. 2018. 'Analyse du discours de la liminalité: Voyageurs britanniques dans les Alpes au xix$^{e}$ siècle', in Marie-Christine Fourny and Stéphane Gal (eds), *Montagne et liminalité: Les manifestations alpines de l'entre-deux XVIe–XXIe siècles*. Grenoble: PUG, pp. 45–60.

Ounoughi, Samia, Emmanuelle Peraldo and Anne-Florence Quaireau (eds). 2025. *Twenty-First-Century Perspectives on British Travel Writing: Decentering Epistemologies*. Abingdon: Routledge.

Pettinger, Alasdair. 2013. 'Gourdes and Dollars: How Travel Writers Spend Money', in Corinne Fowler, Charles Forsdick and Ludmilla Kostova (eds), *Travel and Ethics*. Abingdon: Routledge, pp. 150–64.

Pépy, Emilie-Anne. 2018. 'Dresser les montagnes contre la peste, la mise en place d'un cordon sanitaire dans le duché d'Aoste au XVIIIe siècle', in Marie-Christine Fourny and

Stéphane Gal (eds), *Montagne et liminalité: Les manifestations alpines de l'entre-deux XVIe–XXIe siècles*. Grenoble: PUG, pp. 191–207.
Ricœur, Paul. 1983–85. *Temps et récit*. 3 volumes. Paris: Seuil.
Schafer, R. Murray. 1977. *The Soundscape: The Tuning of the World*. New York: Knopf.
Strabo. 1917. *Geography, Vol. I*. Loeb Classical Library. Cambridge, MA: Harvard University Press.
Van Gennep, Arnold. 1969 (1906). *Rites de Passage*. Paris: Mouton.
WordReference. 2025. 'Immobility'. Retrieved 28 June 2025 from https://www.wordreference.com/definition/immobility.
Wolff, Francis. 1997. *Dire le monde*. Paris: Presses Universitaires de France.

# Index

www.ingramcontent.com/pod-product-compliance
Lightning Source LLC
LaVergne TN
LVHW011332040326
832673LV00002B/39

* 9 7 8 1 8 3 6 9 5 3 5 5 5 *

Doris Stöhr-Mäschl

# Blitzentspannung für Grundschulkinder

## Schnelle Übungen für kurze Auszeiten

Verlag an der Ruhr

# Impressum

**Titel**

Blitzentspannung für Grundschulkinder

*Schnelle Übungen für kurze Auszeiten*

**Autorin**

Doris Stöhr-Mäschl

**Illustrationen**

Eva Spanjardt

**Titelbildmotiv**

**© *pressmaster – Fotolia.com***

**Druck**

AZ Druck und Datentechnik GmbH, Kempten, DE

**Verlag an der Ruhr**

Mülheim an der Ruhr

www.verlagruhr.de

**Geeignet für die Klassen 1–4**



**ISBN 978-3-8346-3686-7**

# Inhaltsverzeichnis

# Vorwort

*„In der Ruhe liegt die Kraft."*

Diese alte Weisheit wird im Alltag oft zitiert, wenn es darum geht, anstehende Aufgaben zielführend zu erfüllen, ohne dabei den „Kopf" und dadurch die Konzentration zu verlieren.

Blicken wir in die Klassenzimmer unserer Grundschulen, finden wir immer häufiger eine für Lehrer und Schüler oft gleichermaßen unbefriedigende Unterrichtssituation vor.

Störende, zappelnde, nicht ausgeschlafene, von Unruhe getriebene Schüler können dem Unterrichtsgeschehen meist nur mit viel Nachdruck und zeitintensiven Wiederholungen folgen.

Lehrer stehen nicht selten vor der Frage, wie sie Wissensvermittlung mit bzw. trotz des wachsenden defizitären Sozialverhaltens einiger Schüler in Einklang bringen können. Meist wird den Lehrern dabei sehr viel Energie und Einfallsreichtum abverlangt, um den Unterrichtsstoff dem oft gravierend differierenden Schülerpotenzial anzupassen.

Die Durchführung von kindgerecht entwickelten Entspannungsangeboten stellt heutzutage einen immer wichtiger werdenden Beitrag zur Gesundheitsförderung dar und könnte bei regelmäßiger Durchführung den Lernerfolg maßgeblich steigern.

Gibt man Kindern immer wieder die Möglichkeit, sich geistig und körperlich zu erholen, geraten sie seltener in den für Körper und Geist anstrengenden Zustand höchster körperlicher Erregung und Alarmbereitschaft. Dabei kann die konsequente Lockerung der Muskulatur in Verbindung mit bewusster Atmung helfen, sich besser zu konzentrieren, um auf Dauer länger leistungsfähig zu bleiben.

Als Lehrer kennen Sie sicher die mentalen Auszeiten, in denen der Schüler mehr oder weniger träumend dem Unterricht folgt. In dieser für den Menschen so wichtigen „Reinigungsphase" verlangsamt sich die Gehirntätigkeit und ermöglicht, sich in einem Zustand mittlerer Entspannung zu regenerieren.

Der wohltuende Wachzustand – auch vielen bekannt als Tiefenentspannung oder „autogenes Training" – wird auch als Kreativphase beschrieben, in der Informationen bestmöglichst verarbeitet werden und sich unser Körperempfinden spürbar und nachhaltig verbessert.

Durch den rituellen und regelmäßigen Einsatz der in diesem Buch vorgestellten Übungen zur Blitzentspannung werden Rahmenbedingungen geschaffen, die es unseren Kindern erlauben, aus dem „Funktionsmodus" auszubrechen. So lernen sie gezielt, ihren Körper und ihre Körpersignale wahrzunehmen, und entwickeln schon sehr früh ein gesundes, körperliches Bewusstsein.

In meiner täglichen Zusammenarbeit mit Schülern, Lehrern, Psychologen und Kinderärzten stelle ich immer wieder fest, wie wichtig gerade im Zeitalter wachsenden Medienkonsums und wachsender Unsicherheiten im erzieherischen Umgang mit Kindern eine wohltuende „Blitzentspannung" aller am Erziehungsprozess Beteiligten geworden ist. Lehrer und Schüler profitieren gleichermaßen von den kurzen Entschleunigungsübungen, die als wohltuende und konzentrationsfördernde Gedankenstopps, als achtsame Körperwahrnehmung oder als fantasievolles „Umschalten" ein entspanntes Durchatmen im Unterrichtsalltag erlauben.

Dieses Buch vermittelt in übersichtlicher und leicht nachvollziehbarer Form, wie Sie ohne großen Aufwand größere Gelassenheit erreichen und die Lernmotivation der Schüler durch konzentrations- und ruhefördernde Übungen steigern können.

Ich wünsche Ihnen und Ihren Schülern viel Erfolg und nachhaltige Entspannung bei der Umsetzung der Übungen.

An dieser Stelle möchte ich mich auch noch bei meiner großartigen Familie und bei den lieben Schülern, Eltern und Lehrern für das ideenvolle und lustige Mitwirken beim Erproben der Übungen bedanken.

Mein besonderer Dank gilt dir, liebe Konny, für alles, was ich in dieser Zeit an Hilfen und persönlicher Unterstützung durch dich erfahren durfte.

*Doris Stöhr-Mäschl*

# Zum Einsatz dieses Buches

## Aufbau

Alle Übungen zur **Blitzentspannung** sind speziell für Grundschulkinder im Alter von sechs bis zehn Jahren konzipiert, können aber auch ohne Weiteres mit Vorschulkindern durchgeführt werden.

Das Buch ist in vier Kapitel unterteilt. **Im ersten Kapitel** finden Sie ein in der Praxis erprobtes und schnell erlernbares **6-Stufen-Basisprogramm**, das, aufeinander aufgebaut, zu einer dauerhaften Grundentspannung beiträgt.

Einmal erlernt, kann das Basisprogramm in ca. sechs Minuten schnell und wirksam durchgeführt werden. Als Muskelentspannungsübung, verknüpft mit einer bewussten Atemlenkung, sorgt das 6-stufige Basisprogramm für innere Ruhe und fördert dadurch die Konzentrationsfähigkeit.

Die einzelnen Übungen beinhalten kleine mentale **Motivationsformeln**, sogenannte Suggestivsätze, die den Kindern den erwünschten Übungszustand vermitteln sollen. Die mentale Einstimmung führt die Kinder bildhaft und thematisch an die jeweilige Übung heran und unterstützt so die kindliche Vorstellungskraft.

In der übersichtlichen **Schautafel** (siehe Seite 14/15) finden Sie die Kurzanleitung des Basisprogramms. Die Kopiervorlagen geben den Schülern eine zusätzliche Hilfestellung zur raschen Umsetzung und Verinnerlichung der Übung. Das Basisprogramm kann sowohl **im Liegen, im Stehen und auch im Sitzen** durchgeführt werden.

Der Einfachheit halber ist im Überblick und in der Darstellung die sitzende Form gewählt. Um eine zu große Unruhe beim Positionswechsel, z. B. durch das Verschieben der Stühle, einen zu großen Geräuschepegel beim Aufstehen und beim Hinlegen usw., zu vermeiden, sollten Sie sich deshalb vor Übungsbeginn auf eine allgemeingültige Position einigen.

Haben die Schüler das Basisprogramm nach einer kurzen Übungseinweisung in sechs aufeinanderfolgenden Einheiten erlernt, können Sie die Gesamtübung in der Regel in ca. sechs Minuten durchführen.

**In den Kapiteln 2 bis 4** werden weitere, in der Praxis erprobte und von Kindern gern angenommene Übungen zur Blitzentspannung vorgestellt. Die Kurzübungen aus dem sensorischen, imaginativen und dem auf Achtsamkeit basierenden Bereich können schnell und ohne großen Aufwand im Schulalltag durchgeführt werden.

Die **sensorischen Übungen** lenken die Aufmerksamkeit der Schüler konzentrations- und stillefördernd in die Welt der Sinne und sorgen für eine wohltuende körperliche Entspannung.

Die **imaginativen Übungen** erzeugen durch das Anstupsen der kindlichen Fantasie wohltuende Entspannungsmomente und bringen die Kinder in die für die Wissensaufnahme so wichtige Regenerationphase.

Die **auf Achtsamkeit basierenden Übungen** lenken die Wahrnehmung gezielt in den für die Achtsamkeit so bedeutenden Moment der Gegenwart und schulen die gesunde Atemwahrnehmung, die Wahrnehmung der Gedanken und das positive Denken.

Die **mentale Einführung** aller im Buch vorgestellten Übungen schafft in Sekundenschnelle einen thematischen und bildhaften Zugang und trägt somit auch zur Entspannungsmotivation besonders unruhiger Schüler bei.

Die gepunkteten Linien in den Texten kennzeichnen kleine Sprechpausen, die den Kindern die Möglichkeit zum Nachspüren oder Träumen geben. Die Länge der Pausen können Sie der Entspannungsbereitschaft Ihrer Schüler und Ihrem zeitlichen Rahmen anpassen.

## Zur Umsetzung der Übungen

- Beginnen Sie alle Übungen stets mit einem **kleinen Einstiegsritual**, das kann z. B. das Lüften des Zimmers, eine brennende Kerze oder ein guter Duft sein. **Wohlriechende Düfte**, wie z. B. Zitrusduft, werden von den Schülern als besonders inspirierend empfunden und bringen zudem eine **harmonische Atmosphäre** in das Klassenzimmer.
- Eine **ruhige und angenehm tiefe Stimme sowie leise und entspannende Musik** im Hintergrund erleichtern den Übungseinstieg und verstärken die Übungsbereitschaft Ihrer Schüler.

- **Positive Gedanken** stärken die innere Einstellung zu einer Tätigkeit. Die Übungen im 6-Stufen-Programm beinhalten deshalb kurze Motivationsformeln bzw. Suggestivsätze.

- Der tiefere Sinn jeder Entspannung liegt in der **Körperwahrnehmung**. Es ist deshalb wichtig, den Kindern Raum und Zeit zu geben, ihre Körperempfindungen zu spüren. Bei unruhigen und leicht ablenkbaren Schülern können Sie durch Ihre **Nähe** mit leichten **Berührungen der zu entspannenden Körperteile** die Körperwahrnehmung lenken und unterstützen.

- Während der Durchführung von Entspannungsübungen können in sehr seltenen Fällen für das Kind **unangenehme und ungewohnte Empfindungen,** wie z. B. Taubheitsgefühle in Händen und Beinen, vermehrtes Muskelzucken und leichter Schwindel, aufkommen. Diese sind jedoch vollkommen harmlos und treten nach mehrmaligem Üben nicht mehr auf.

- Vermeiden Sie im Vorfeld, unruhige und verhaltensauffällige Kinder mit „Pausenabsitzen auf der Ruhebank" zu bestrafen. Ein **„auspowerndes Sanktionieren"**, wie z. B. Strecken laufen, Hüpf- oder Kletteraufgaben, fördern dagegen die erwünschte Ruhebereitschaft von „Zappelschülern" wesentlich effektiver.

- In der Entspannung gibt es **weder richtig noch falsch**. Alles, was guttut, dient der Entspannung. Führen Sie die **Übungen stets bewertungsfrei** und **ohne Vergleiche** durch. Ruhe und Entspannung wird von uns Menschen ganz unterschiedlich wahrgenommen, angenommen und empfunden.

- Störquellen behindern den Übungsablauf und lenken stets ab. Suchen Sie nach Zeiten, zu denen ein **störungsfreier Durchlauf** der Übungen garantiert ist. Spürbar deutlicher wird Entspannung **nach intensiver Bewegung** wahrgenommen. Falls möglich, empfehle ich deshalb, die Übungen zu Beginn des Unterrichtes oder im Anschluss an die Pausen durchzuführen. Dadurch lernen die Kinder, ihren Körper schneller runterzufahren und die **innere Ruhe als spürbar und positiv** zu akzeptieren.

- Ein gutes **Vorbild** motiviert die Schüler zur Nachahmung. Wer als Lehrer selbst die Ruhe spürt, wird sie auch überzeugter weitergeben. Machen Sie sich vorher selbst **mit den Übungen** vertraut und vermitteln Sie die Blitzentspannung als kleines Experiment. Damit wecken Sie die Neugier Ihrer Schüler. Eine bewusst langsame

Anleitung gibt den Schülen die Möglichkeit, sich in ihrem Tempo in die Übung reinzufinden.

- Erarbeiten Sie mit den Kindern Regeln zum Fair Play. Gegenseitige Rücksichtnahme ohne Auslachen sollte selbstverständlich sein!
- Alle im Buch aufgeführten Übungen können wesentlich zur Beruhigung der Kinder im Unterricht beitragen. Versuchen Sie trotz allem, die Ursachen für Unruhe und Hyperaktivität Ihrer Schüler herauszufinden. Manche Ursachen für körperliche Unruhe, Zappeln, Unkonzentriertheit usw. können auch durch eine zunehmende „Terminhetze" der Kinder, zu langen und unbeaufsichtigten Medienkonsum, schlechtes Ein- und/oder Durchschlafen, schlecht verdauliche und späte Nahrungsaufnahme, Hunger, ständiges Vergleichen mit Mitschülern, „Förderwahnsinn" am Nachmittag oder familiäre Probleme hervorgerufen werden. Regelmäßig integrierte Entspannungsübungen mit anschließender Kurzreflexion können Ihren Schülern beim Artikulieren und bei der Wahrnehmung ihrer Probleme helfen.
- Emotional belastete und hippelige Kinder können fremdbestimmte Entspannungsvorgaben als bedrohlich und unangenehm empfinden. Geben Sie diesen Kindern die Möglichkeit, sich langsam und selbstbestimmt an die Entspannung heranzutasten, z. B. durch das aktive Einbeziehen in die Rolle des Anleiters.
- Aufgrund unterschiedlicher taktiler Sensibilität können Kinder sehr verschieden auf Berührungen reagieren. Die Bandbreite reicht von extrem kitzelig bis angenehm entspannt. Aus diesem Grund habe ich bewusst auf Übungen durch Fremdberührung verzichtet und stattdessen die Selbstberührung in den Vordergrund gestellt.
- Entspannungsübungen können zu einem großen Teil zur Gesundheitsentwicklung unserer Kinder beitragen. Trotz der vielfach belegten und wissenschaftlich nachgewiesenen Erfolge von Entspannungsübungen stellen sie jedoch kein Allheilmittel bei schulischen Lernproblemen und defizitärem Verhalten dar.
- Bei chronischem Asthma, bei angeborenen Herz- und/oder Kreislauf-Erkrankungen, bei chronischen Magen-Darm-Problemen und bei Anfallserkrankungen, wie Epilepsi, sollten Sie vor der Durchführung Rücksprache mit den Eltern halten.

# Das 6-Stunden-Entspannungsprogramm

# Einführung

Das 6-Stufen-Programm der Blitzentspannung wurde von mir ursprünglich als Einsteigerübung für Familien im Bereich des Stressmanagements entwickelt. Damit konnte ich besonders entspannungsunfähige Kinder bzw. Familien spielerisch und körperorientiert an eine Tiefenentspannung heranführen. Durch das Einbeziehen verschiedener, besonders im Schulalltag angestrengter Muskelgruppen im Kontext einer bewusst durchgeführten Atmung erzielte ich bei extrem angespannten Grundschülern einen nachhaltig wirksamen und schnellen Entspannungserfolg.

Die sehr einfach und bildhaft dargestellten Übungen, hierzu zählten der **„Bauchballon“** für einen entspannten Bauch- und Darmbereich, der **„Kletterfrosch“** für entspannte Finger- und Handmuskeln, der **„Ruderer“** für entspannte Arm- und Schultermuskeln, die **„Raupe“** für die Durchblutung der Fuß- und Beinmuskulatur, der **„Adlerflug“** für eine wohltuende Ganzkörperentspannung, werden mit dem **„Körperdetektiv“**, unserer mentalen Schlussentspannung, abgeschlossen.

Die leicht verständlichen Übungen bauen aufeinander auf und bringen nach einer kurzen Erlernphase eine wohltuende Blitzentspannung für Körper und Geist. Hängen Sie dazu an jedem neuen Übungstag die nächste Übung an.

Ziel ist es, den Übungsablauf so zu verinnerlichen, dass jeder Schüler nur durch Vorzeigen der beiliegenden Kopiervorlagen und durch Vorsagen des Suggestivsatzes die Übung gedanklich und körperlich nachvollziehen kann. Ein über mehrere Wochen täglich durchgeführtes Basisprogramm wird so von den Schülern als ein willkommenes Ruheritual angenommen.

Die Übungen werden von einer bewussten und langsamen Atmung begleitet. Unsere Atemzüge, im entspannten Zustand ca. 23 000 am Tag, sind sozusagen der Motor, der unseren Energie-Kreislauf in Bewegung hält und somit wichtige Körperfunktionen, wie z. B. auch die Verdauung oder das Denken, erst möglich macht.

Deshalb spielt die Atmung eine wesentliche Rolle bei der Durchführung der Entspannungsübungen. Die Atmung ist unser verlässlichstes Instrument, wenn es darum geht, unseren Körper zu beruhigen, ihn bewusst zu spüren und unsere körperliche und geistige Leistungsfähigkeit zu stabilisieren oder auch zu optimieren.

Bei jedem Atemzug dehnen sich unsere beiden Lungenflügel im Brustkorb aus und benötigen viel Platz. Das Zwerchfell, unser zentraler Atemmuskel, der die eingeatmete Luft in die Lunge schleust, schiebt beim Einatmen unsere Organe in den Bauchraum und sorgt so für den benötigten Platz unserer luftgefüllten Lungenflügel.

Damit unsere Bauchorgane durch den Druck von oben nicht zusammengequetscht werden, was wir häufig als Bauchschmerzen wahrnehmen, ist es nötig, die Bauchdecke beim Einatmen weit nach außen zu wölben, um Platz für unsere Organe (Magen, Darm, Niere, Herz, Leber) zu schaffen. Die bewusste Bauchatmung belebt zusätzlich unseren Blut- und Herzkreislauf, baut Stress und emotionale Spannungen ab und erhöht so die Leistungsfähigkeit unseres ganzen Organismus.

Beim Ausatmen entspannt und lockert sich unsere Atem- und Bauchmuskulatur wieder, die Bauchdecke verliert ihre Spannung, unsere Muskulatur bereitet den nächsten Atemzug vor. Zwischen Ein- und Ausatmen sorgt eine kleine Atempause für Entspannung.

Das intensive Ausatmen setzt zusätzlich eine Dosis Kalzium frei, das einer Verengung der Nerven entgegenwirkt. Somit lassen sich Kopfschmerzen, Konzentrationsprobleme, Müdigkeit oder Unruhe gut „wegatmen".

# Übersicht

| | Der Bauchballon | Der Kletterfrosch | Der Ruderer |
|---|---|---|---|
| **Mentale Einstimmung –visuelle Vorstellung** | Stelle dir deinen Bauch als Luftballon vor. | Stelle dir einen Frosch vor, der an einer glatten Fläche hochklettert. | Stelle dir vor, wie du mit beiden Armen ruderst. |
| **Vorbereitung** | • Entspanntes, zurückgelehntes Sitzen<br>• Hände liegen auf dem Bauchnabel<br>• Füße sind geerdet, berühren also mit der gesamten Sohle den Boden<br>• Augen sind geschlossen oder fixieren einen festen Punkt | • Aufrechtes Sitzen<br>• Hände liegen aufgefächert auf dem Tisch<br>• Füße sind geerdet<br>• Augen sind geschlossen oder fixiert | • Aufrechtes Sitzen auf der Stuhlkante<br>• U-Haltung beider Arme in Brusthöhe<br>• Augen sind geschlossen oder fixiert |
| **Durchführung** | „Atme langsam durch die Nase ein und lenke die Luft in den Bauchballon. Atme aus und lasse deinen Bauchballon wieder kleiner werden." | „Atme langsam ein. Deine Finger ‚saugen sich auf dem Tisch an'. Atme aus und löse deine Saugnäpfe." | „Atme langsam ein und ziehe die Ruder seitlich an deine Brust. Atme aus und lasse die Ruder wieder sanft nach vorn gleiten." |
| **Rücknahme** | Augen öffnen, Arme und Beine strecken und ausschütteln | Augen öffnen, Arme und Hände ausschütteln | Augen öffnen, Arme und Beine kräftig ausschütteln |
| **Formel** | „Ich atme frei und bin entspannt!" | „Meine Finger sind locker und entspannt!" | „Meine Arme sind ganz entspannt!" |

| | Die Raupe | Der Adlerflug | Der Körperdetektiv |
|---|---|---|---|
| **Mentale Einstimmung – visuelle Vorstellung** | Stelle dir eine Raupe vor, die sich langsam vorwärtsbewegt. | Stelle dir einen fliegenden Adler mit seinen riesigen Flügeln vor. | Du bist heute ein Detektiv, der seinen eigenen Körper erforscht. |
| **Vorbereitung** | • Aufrechtes Sitzen auf der Stuhlkante<br>• Fußsohlen sind ganzflächig geerdet<br>• Augen sind geschlossen oder fixieren einen Punkt | • Aufrechtes Sitzen, Füße sind geerdet<br>• „Armflügel" sind weit gespannt<br>• Augen sind geschlossen oder fixiert | • Entspanntes, zurückgelehntes Sitzen<br>• Füße sind geerdet<br>• Arme hängen entspannt nach unten<br>• Augen sind geschlossen oder fixiert |
| **Durchführung** | „Atme langsam ein und ziehe dabei deine Füße mit den Zehen nach vorn." | „Atme langsam ein und schwinge deine Arme nach oben, atme aus und lasse die Arme absinken." | „Gedankenwanderung durch den Körper:<br>Finger, Arme, Schultern, Kopf, Rücken, Bauch, Beine, Füße." |
| **Rücknahme** | Augen öffnen, Körper strecken und ausschütteln | Augen öffnen, Arme kräftig ausschütteln | Augen öffnen, ganzen Körper strecken und ausschütteln |
| **Formel** | „Meine Füße sind ganz entspannt!" | „Mein ganzer Körper ist ganz entspannt!" | „Ich bin ganz ruhig und entspannt!" |

# Stufe 1: Der Bauchballon

## Ziel der Übung

- Der Bauchballon sorgt für **Entspannung im Bauchbereich**. Durch lange Sitzperioden während des Unterrichts und das häufig zu beobachtende Unterdrücken des Stuhl- und Harndrangs treten bei Schulkindern oftmals Blähungen und Bauchschmerzen auf. Diese Übung trägt zur **Entkrampfung des Magen- und Darmbereichs** bei.

## Das sollten Sie wissen

- Kinder neigen im Allgemeinen dazu, Übungen schnell und unkontrolliert durchzuführen. Um einem Hyperventilieren, einer zu schnellen Atmung und einem Schwindelgefühl vorzubeugen, ist es ratsam, mit den Kindern die **langsame und rhythmisierte Bauchatmung** immer wieder zu trainieren. Lassen Sie die Kinder vor der ersten Atemübung erst mal kräftig bei offenen Fenstern gähnen. Das intensive Gähnen gilt als die **natürlichste Einatemübung**, da die Skelettmuskulatur wohltuend gedehnt, der Organismus gut durchblutet und das häufig angespannte Kiefergelenk gelockert wird.
- Magen- und Darmgeräusche gelten bei der Entspannung als normal und sollten daher nicht unnötig thematisiert werden.
- Zeigen Sie den Kindern, wie eine bewusste und entschleunigte Atmung bei Ihnen aussieht. **Atmen Sie vor!**

## Mentale Einstimmung

„Dein Körper atmet von ganz allein, du brauchst dafür gar nichts zu machen, stimmt's? Dein Gehirn kümmert sich ständig um die Versorgung deines Körpers mit Atemluft. Manche Menschen haben eine Atemwegserkrankung und müssen das gleichmäßige Atmen wieder lernen, z. B. bei Asthma. Auch für gesunde Menschen ist es gut, zu lernen, wie man ruhig und gesund atmet.

Hast du schon mal mit deiner Nase einen Luftballon aufgeblasen? Um dies auszuprobieren, brauchst du aber keinen bunten Luftballon aus Gummi, sondern nur deinen Körper. Deinen Luftballon trägst du nämlich immer mit dir herum.

Kannst du dir vorstellen, welcher Körperteil sich wie ein Ballon aufblasen lässt? – Richtig, es ist dein Bauch. Probiere es doch mal aus!"

## Durchführung

„Setze dich locker und angelehnt auf deinen Stuhl ...... Lege beide Hände auf deinem Bauchnabel übereinander ...... Deine Fußsohlen stehen fest auf dem Boden ...... Deine Augen sind geschlossen oder auf einen festen Punkt vor dir gerichtet ...... Atme langsam durch deine Nase ein und wieder aus ...... Spüre das Auf und Ab deiner Bauchdecke ...... Atme wieder ein und halte kurz die Luft an ...... Atme schnell aus ...... Dein Bauch wird schnell ganz klein ...... Blase deinen ‚Bauchballon' mit deiner Nase langsam auf ...... **‚Ich atme frei und bin entspannt!'** ...... Lasse die verbrauchte Atemluft durch deine Lippen langsam entweichen ...... Dein ‚Bauchballon' wird kleiner und kleiner ...... Dein Körper atmet von selbst ein ...... Fülle deinen Bauchballon noch einige Male mit frischer Atemluft ......'

## Rücknahme

Die folgende Rücknahme findet hier nur statt, wenn Sie das 6-Stufen-Programm mit dieser Übung beenden wollen:
„Strecke und rekele dich nun so gut du kannst ...... Öffne deine Augen ...... Komme mit deinen Gedanken wieder in das Klassenzimmer zurück!"

## Abschluss

„Wie fühlen sich dein ‚Bauchballon' und der ganze Körper nun an?"

# Der Bauchballon

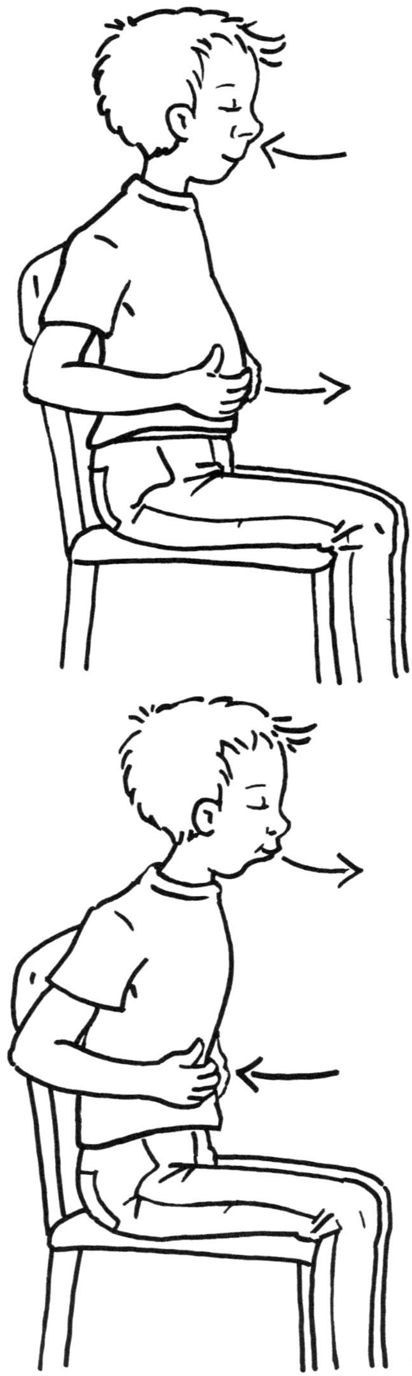

© Eva Spanjardt

# Stufe 2: Der Kletterfrosch

## Ziel der Übung

- Das abschließende Ausschütteln der Finger und Hände sorgt für eine **angenehme Durchblutung**. Die Finger-Fächer-Übung **unterstützt die Feinmotorik** und lockert die Finger- und Handmuskulatur. Die Hände ermüden weniger schnell, die **Handschrift bleibt leserlicher**.

## Das sollten Sie wissen

- **Erklären** Sie bei Bedarf den Begriff „Saugnäpfe".
- Der Kletterfrosch kann selbstverständlich auch seine Fuß-Saugnäpfe am Boden festsaugen lassen. Dies ist aber eher eine **Sommerübung**.

## Mentale Einstimmung

„Es gibt Frösche, die können mit ihren Saugnäpfen problemlos an glatten Flächen, wie Blättern oder Fensterscheiben, senkrecht nach oben laufen. Hast du schon mal einen Frosch beobachtet? Willst du mal ausprobieren, wie sich deine ‚Saugnäpfe' an deinem Tisch ansaugen?"

## Durchführung

„Setze dich aufrecht auf deinen Stuhl ...... Deine Fußsohlen stehen fest auf dem Boden ...... Beide Hände liegen aufgefächert vor dir auf dem Tisch ...... Deine Finger sind weit auseinander gespreizt ...... Atme zwei Atemzüge tief ein – und wieder aus ...... Deine Augen sind entweder geschlossen oder auf einen fixen Punkt in der Ferne gerichtet ...... Atme nun tief und langsam ein und drücke dabei deine Finger wie Saugnäpfe einzeln und kräftig in den Tisch ...... Atme aus und löse deine Saugnäpfe ...... Sage dir in Gedanken deine Entspannungsformel vor: **‚Meine Finger sind locker und entspannt!'** ...... Atme weiter und wiederhole die Übung noch 3-mal ...... Spüre die Entspannung in deinen Fingern ...... Schüttel deine ‚Froschhände' gut aus!"

## Rücknahme

Die folgende Rücknahme findet hier nur statt, wenn Sie das 6-Stufen-Programm mit dieser Übung beenden wollen:
„Strecke und rekele dich nun so gut du kannst ...... Öffne deine Augen und komme mit deinen Gedanken wieder in den Klassenraum zurück!"

## Abschluss

„Wie fühlen sich deine Hände/Finger an? Kannst du die Wärme in deinen Händen spüren?"

# Der Kletterfrosch

# Stufe 3: Der Ruderer

## Ziel der Übung

- Diese Übung mobilisiert den **gesamten Oberkörper** und kräftigt den **Arm- und Schulterbereich.**

## Das sollten Sie wissen

- Achten Sie auf eine **langsame** und **sorgfältige Durchführung** der Übung.
- **Beide Arme bewegen sich synchron** zum Ein- und Ausatmen.
- Beim Ausstrecken der Arme sind die **Hände gelockert**.
- Beide Ellenbogen zeigen beim Heranziehen der Ruder in Brusthöhe weit nach hinten. Damit wird der **Brustbereich geweitet**, die Luft kann gut einfließen.
- **Langsame, kontrollierte Bewegungen** verstärken die anschließende Entspannung.

## Mentale Einstimmung

„Bist du schon mal mit einem Ruderboot mitgefahren oder hast sogar selbst schon mal ein Boot gerudert? Dafür braucht man ganz schön viel Kraft, stimmt's? Heute nimmst du mit deiner Klasse an einem Ruderwettbewerb teil. Jeder rudert mit beiden Händen sein eigenes Boot. Das langsame Durchziehen der Ruder kostet ganz schön Kraft und viel Atem. Nur wer langsam und kraftvoll rudert, wird auch gewinnen."

## Durchführung

„Stelle dir eine Ruderbank in einem Boot vor ...... Setze dich aufrecht darauf und nimm mit beiden Händen die schweren Rudergriffe in die Hand ...... Beide Ellenbogen sind seitlich in Brusthöhe angewinkelt ...... Deine Fußsohlen sind ganzflächig abgestellt ...... Atme ein und ziehe die Ruder ganz nahe seitlich an deine Brust ...... Atme aus und lasse die Rudergriffe locker nach vorn gleiten ...... Sage dir beim Ausatmen in Gedanken deine Entspannungsformel vor ...... **‚Meine Arme sind ganz entspannt.'** ...... Wiederhole die Ruderbewegung und die Entspannungsformel noch ein paar Atemzüge lang ...... Lasse anschließend die Arme locker ...... Spüre die Entspannung in deinen Armen."

## Rücknahme

Die folgende Rücknahme findet hier nur statt, wenn Sie das 6-Stufen-Programm mit dieser Übung beenden wollen:

„Öffne deine Augen und lasse deine ‚Ruder-Arme' locker nach unten fallen."

## Abschluss

„Welche Muskeln kannst du nun spüren?"

# Der Ruderer

© Eva Spanjardt

# Stufe 4: Die Raupe

## Ziel der Übung

- Bei dieser Übung wird der gesamte **Fuß- und Wadenmuskel** erst angespannt und schließlich durch das Nachspüren entspannt. Die Muskulatur wird dabei wohltuend durchblutet.

## Das sollten Sie wissen

- Diese Übung sollte, **wenn möglich, ohne Schuhe** durchgeführt werden, damit man die Bewegungen der Füße gut spüren kann.
- Üben Sie mit den Kindern **die Raupenbewegung** vorher kurz ein.
- Achten Sie auf eine **langsame Fußführung**.
- Beim **Raupen-Rückwärtsgang** kann man die Muskelanspannung deutlicher spüren. Heben Sie dabei die Fersen vor dem Aufsetzen nach oben, damit werden die Wadenmuskeln zusätzlich angespannt.
- Achten Sie auf eine **aufrechte Sitzposition**.
- **Variante im Stehen:** Lassen Sie Ihre Schüler mal im **Raupentempo** stehend durch die Klasse „kriechen".

## Mentale Einstimmung

„Hast du schon mal eine Raupe bei der Fortbewegung beobachtet? Bei genauerem Hinsehen können wir entdecken, dass Raupen sich erst mal lange ausstrecken und dann eine Brücke bilden, indem sie die kräftigen hinteren Nachschieberbeine direkt hinter die vorderen Bauchbeine stellen. So schieben sie sich in ihrem Tempo langsam durch das Gelände. Willst du auch mal ausprobieren, wie sich diese Fortbewegung an deinen Füßen anfühlt?"

## Durchführung

„Setze dich aufrecht auf die vordere Stuhlkante ...... Deine Fußsohlen stehen platt unter der Sitzfläche auf dem Boden ...... Deine Arme hängen seitlich entspannt neben dem Körper nach unten ...... Deine Augen sind geschlossen oder auf einen Punkt fixiert ...... Atme ein ...... Mache eine Zehenfaust ...... Ziehe deine Zehen mit langsamen Atemzügen wie eine Raupe nach vorn ...... Atme aus und lockere die Zehen ...... Sage dir in Gedanken deine Entspannungsformel vor ...... **‚Meine Füße sind ganz entspannt'** ...... Wiederhole die Übung, bis deine Beine gestreckt sind ...... Nun kommt der ‚Raupenrückwärtsgang' ...... Komme mit langsamen Atemzügen zum Stuhlbein zurück ...... Wiederhole deine Entspannungsformel: **‚Meine Füße sind ganz entspannt'** ......"

## Rücknahme

Die folgende Rücknahme findet hier nur statt, wenn Sie das 6-Stufen-Programm mit dieser Übung <u>beenden</u> wollen:

„Schüttel und lockere deine Füße und spüre anschließend, wie entspannt sie geworden sind."

## Abschluss

„Wie fühlen sich deine Füße und Beine an? Kannst du deine Füße/Beine nun deutlicher spüren als vorher?"

## Die Raupe

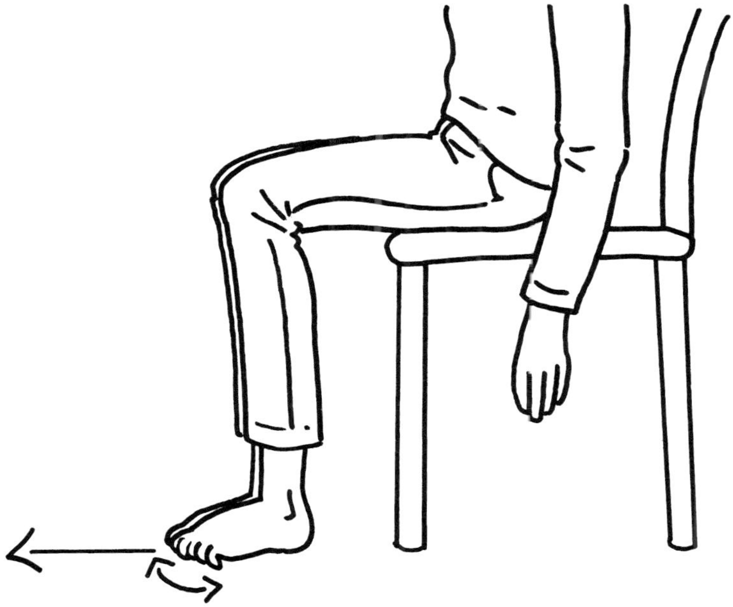

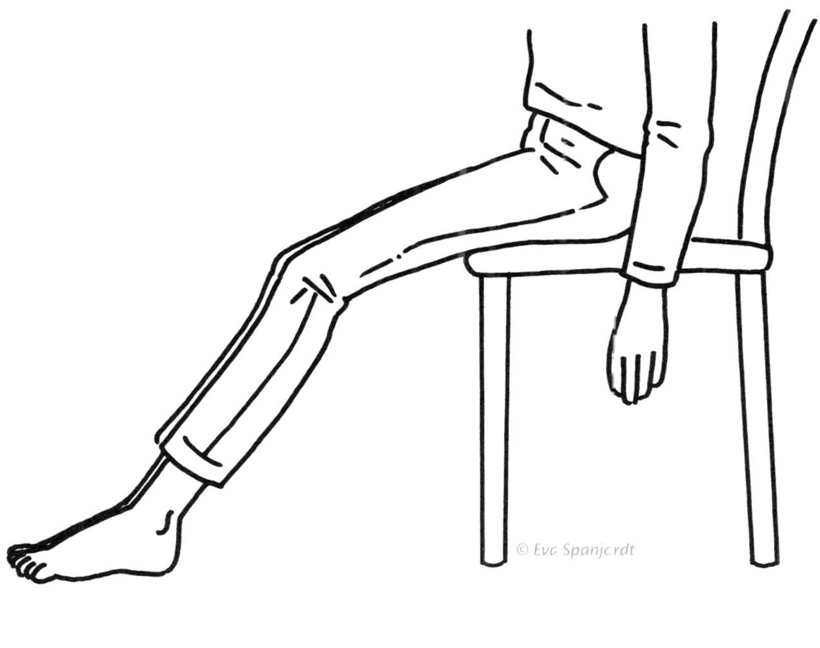

# Stufe 5: Der Adlerflug

## Ziel der Übung

- Diese Übung lockert die **Arm-, Schulter-, Brust- und Rückenmuskeln**.

## Das sollten Sie wissen

- Achten Sie auf genügend **Sicherheitsabstand** zum Nebenmann.
- Kinder neigen zu einer schnellen und hastigen Durchführung. **Sprechen und atmen Sie deshalb entspannt** den Rhythmus vor.
- Probieren Sie die Übung auch mal in **extremer Zeitlupe und im Zeitraffer** aus. Dadurch kommen die Schüler auch mal außer Atem und können anschließend die Muskeln intensiver spüren.

## Mentale Einstimmung

„Der größte Greifvogel bei uns in Deutschland ist der Steinadler. Er wird wegen seiner großen Flügelspannweite von über zwei Metern auch ‚König der Lüfte' genannt. Er kann seine Beute mit seinen Adleraugen auch noch aus großer Höhe auf dem Boden erkennen. In den Alpen fühlt sich der Steinadler sehr wohl, hier lebt und nistet er in den Felsspalten. Strecke deine Arme zur Seite und schaue, wie groß deine Armspannweite ist. Versuche, dir vorzustellen, wie du dich mit deinen Armen und der Kraft deines Atems weit nach oben in die Lüfte schwingst. Du wirst staunen, was du von hier oben alles erkennen kannst."

## Durchführung

„Du sitzt mit geradem Rücken auf der vorderen Stuhlkante …… Beide Arme hängen locker neben deinem Körper nach unten …… Die Handflächen schauen nach außen …… Deine Fußsohlen berühren ganzflächig den Boden …… Deine Augen sind geschlossen …… Atme ein und schwinge deine gestreckten Armen, wie ein Vogel hoch und runter über deinen Kopf und drücke dabei die Luft nach oben ……

Drehe deine Handflächen über dem Kopf nach unten und drücke die Luft nach unten ...... Atme aus und lasse deine Arme langsam nach unten sinken ...... Sage dir in Gedanken deine Entspannungsformel vor: **‚Mein ganzer Körper ist ganz entspannt!'** ...... Breite deine Arme noch mal aus ...... Fliege noch eine Weile mit deinem entspannten Körper über die Landschaft ......"

## Rücknahme

Die folgende Rücknahme findet hier nur statt, wenn Sie das 6-Stufen-Programm mit dieser Übung beenden wollen:

„Öffne deine Augen und lasse deine ‚Adlerflügel' nun locker nach unten fallen."

## Abschluss

„Spürst du die Entspannung in den Armen?"

# Der Adlerflug

# Stufe 6: Der Körperdetektiv

## Ziel der Übung

- Diese Achtsamkeitsübung lenkt den Fokus auf den gesamten Körper, **wirkt beruhigend** vor anstrengenden Aufgaben und unterstützt die Konzentration.

## Das sollten Sie wissen

- Geben Sie den Schülern nach dieser Schlussübung genügend Gelegenheit, über den **Übungserfolg** zu sprechen bzw. Anregungen und Änderungsvorschläge einzubringen.

## Mentale Einstimmung

„Wer bestimmt eigentlich, was du denkst und wie du dich fühlst? Richtig, du bestimmst ganz allein, was du denkst und wie du dich fühlen willst. Damit wir aber immer spüren, wie unser Körper sich anfühlt, müssen wir erst mal lernen, in unseren Körper reinzuhören. Als guter Körperdetektiv kannst du ganz genau die Signale deines Körpers beobachten und spüren. Versuch's doch mal!"

## Durchführung

„Setze dich entspannt auf deinen Stuhl ...... Dein Rücken ist angelehnt ...... Deine Fußsohlen stehen entspannt auf dem Boden ...... Deine Arme hängen seitlich neben dem Stuhl oder liegen entspannt auf den Oberschenkeln ...... Deine Augen sind geschlossen oder auf einen fixen Punkt gerichtet ...... Dein Atem ist ruhig und gleichmäßig ...... Sage dir in Gedanken vor, wie entspannt du bist ...... **‚Ich bin ganz ruhig und entspannt'** ...... Wandere langsam Schritt für Schritt im Rhythmus deines Atems mit deinen Gedanken durch deinen Körper ...... Spüre deine Finger ...... deine Arme ...... deine Schultern ...... deinen Kopf ...... deinen Rücken ...... deinen Bauch ...... deine Beine ...... deine Füße ...... Spüre den schwersten Teil deines Körpers ...... Spüre den wärmsten Teil deines Körpers ...... Spüre den ruhigsten Teil deines Körpers ......"

## Rücknahme

„Öffne Augen und Mund so weit du kannst ...... Strecke deine Arme weit nach außen ...... Hebe deine Beine und schiebe sie weit nach vorn ...... Spreize deine Zehen ...... Schüttel dich jetzt so gut du kannst aus."

## Abschluss

„Wie fühlt sich dein Körper nun an? Hast du den schwersten, wärmsten und ruhigsten Teil deines Körpers spüren können?"

# Der Körperdetektiv

# Sensorische Übungen

# Einführung

Sensorische Übungen vertiefen das Zusammenspiel von taktilen, visuellen, olfaktorischen, akustischen und gustatorischen Reizen, fördern deren bewusste Wahrnehmung und die Wirkung auf das körperliche Wohlbefinden. Sie helfen, ein achtsames Körpergefühl zu entwickeln, körperliche Vorgänge bewusst wahrzunehmen und diese als Grundlage für gesundes Verhalten zu erkennen.

Mithilfe sensorischer Übungen, wie z. B. Atemübungen und Muskelanspannungen, gelangen die Schüler in eine wohltuende und leicht spürbare partielle Entspannung oder Ganzkörperentspannung. Die Kinder lernen, ihre Muskelgruppen zu spüren, ihre muskulären Empfindungen zu deuten und dabei die Konzentration auf die zu entspannende Muskelpartie zu lenken. Das „Loslassen" der Muskeln bringt die erwünschte und eigentliche Entspannung. Dieses von Edmund Jacobson entwickelte Verfahren der „Progressiven Muskelentspannung" wird durch eine bewusst gesteuerte Atmung gestützt.

Unsere bildhaften Kurzübungen vermitteln den Kindern die Fähigkeit, einzelne im Schulalltag sehr angespannte Muskelgruppen spielerisch und anschaulich zu lockern. Die Übungen können in Belastungssituationen, z. B. vor Klassenarbeiten, gezielt im Klassenverband angewendet werden.

# Die Knetmaschine

## Ziel der Übung

- Diese Übung eignet sich zur **Auflockerung und zum Abreagieren** für sehr unruhige Kinder.

## Das sollten Sie wissen

- Achten Sie auf **ausreichend Platz** zum Nebenmann.
- Geben Sie den Kindern Zeit, die **anschließende Ruhe und Entspannung wahrzunehmen.**
- **Entspannungsmusik** kann die Ruhephase positiv unterstützen.

## Mentale Einstimmung

„Bestimmt hast du schon mal deiner Mama beim Kneten eines Teiges zugesehen. Es gibt spezielle Knetmaschinen, die jeden noch so harten Teig ganz mühelos durchkneten. Diese Maschinen können mit einem Schalter mal schneller und mal langsamer geschaltet werden. Heute bist du eine Knetmaschine. Deine Arme sind dabei die Knethaken."

## Durchführung

„Stelle dich mit viel Abstand zu deinen Mitschülern im Raum auf. Bleibe während der Übung möglichst auf diesem Platz stehen.
Schalte die 1. Stufe deiner Knetmaschine ein …… Deine beiden Arme hängen neben dem Körper …… Deine beiden Knethaken (Hände) bewegen sich sehr langsam und kneten in der 1. Stufe einen Teig ……
Schalte in die 2. Stufe …… Die Arme setzen mit ein ……
3. Stufe …… Die Bewegungen werden größer ……
4. Stufe …… Der Oberkörper setzt mit ein ……
5. Stufe …… Der ganze Körper ist in Bewegung …… Die Arme schlängeln sich wild durch die Lüfte und kneten dabei weiter den Teig ……

Schalte in die 4. Stufe zurück ...... Die Bewegungen werden kleiner ......

3. Stufe ...... Nur Arme und Schultern bewegen sich noch ......
2. Stufe ...... Nur die Arme kneten langsam weiter ......
1. Stufe ...... Die Arme bewegen sich ganz ruhig durch den Teig."

## Rücknahme

„Strecke dich zum Abschluss noch mal kräftig in alle Richtungen, setze dich wieder auf deinen Platz und lasse die Ruhe durch deinen Körper fließen."

## Abschluss

„Wo kannst du jetzt die Entspannung im Körper spüren?"

# Die Knetmaschine

© Eva Spanjardt

# Die Sonnenfinsternis

## Ziel der Übung

- Diese Übung ist auch als **Palmieren** bekannt und eignet sich sehr gut nach langen Konzentrationsphasen, aber auch zum Ruhigwerden vor Klassenarbeiten.

## Das sollten Sie wissen

- Geben Sie den Kindern zur Orientierung die Zeit des Reibens und Ruhens vor.
- Die Übung spricht auch Kinder an, die sich nur ungern die Augen zuhalten.
- **Brillenträger** sollten die Brille vor der Übung absetzen.

## Mentale Einstimmung

„Hast du schon mal eine Sonnen- oder Mondfinsternis gesehen oder davon gehört? Bei der Sonnenfinsternis schiebt sich der Mond zwischen Sonne und Erde.
Bei der Mondfinsternis ist die Erde zwischen Sonne und Mond. Heute kannst du mit deinen Händen deine eigene Sonnenfinsternis zaubern."

## Durchführung

„Setze dich aufrecht hin ...... Stelle deine Fußsohlen auf dem Boden ab ...... Reibe deine Handinnenflächen aneinander, bis sie angenehm warm sind ...... Lege beide Hände leicht gewölbt über die geschlossenen Augen, ohne die Augenlider zu berühren ...... Spüre die Wärme deiner Handflächen ...... Atme ruhig und gleichmäßig tief ein ...... Und wieder aus ...... Ein ...... Aus ...... Genieße deine Sonnenfinsternis ......"

## Rücknahme

„Beende jetzt die Übung mit einem langen Strecken und Gähnen."

## Abschluss

„Wie hat sich deine Sonnenfinsternis angefühlt?

# Der Nasenstift

## Ziel der Übung

- Die Übung dient der Auflockerung nach Schreib- und Konzentrationsphasen.

## Das sollten Sie wissen

- Erklären Sie den Kindern vor der Übung, wer Pinocchio ist.
- Sie können Wörter vorgeben, die ein Kind mit dem Nasenstift vormacht. Alle anderen versuchen, das Wort zu erraten, allerdings ist dies dann spiegelverkehrt!
- Achten Sie auf langsame Bewegungen, damit die Muskulatur sich anpassen kann.
- Schreibschrift ist fließender und schonender für die Muskeln als Druckschrift.

## Mentale Einstimmung

„Stelle dir mal vor, du hättest so eine lange Nase wie Pinocchio oder noch viel länger und du könntest sogar damit schreiben. Willst du deinen Nasenstift mal ausprobieren?"

## Durchführung

„Stelle dich gut geerdet an deinen Platz ...... Die Knie sind leicht gebeugt ...... Deine Arme hängen nach unten ...... Dein Kopf ist gerade ...... Dein Blick ist nach vorn gerichtet ...... Atme tief durch die Nase ein ...... Spüre, wie dein Bauch sich nach außen wölbt ...... Atme mit einem Zischgeräusch aus ...... Spüre deine Nasenspitze ...... Sie ist nun dein Stift ...... Schreibe deinen Namen ...... Schreibe die Buchstaben so groß du kannst ......"

## Rücknahme

„Klopfe mit den Fingerspitzen deinen Nacken- und Schulterbereich ...... Schüttel Arme und Beine kräftig aus ...... Beende die Übung mit einem langen Strecken."

## Abschluss

„Spürst du deine Hals- und Nackenmuskeln?"

# Das Augenpendel

## Ziel der Übung

- Diese Übung **trainiert und entspannt müde Augenmuskeln.**

## Das sollten Sie wissen

- Sie können die Übung im **Stehen, Sitzen oder Liegen** machen.
- Achten Sie bei der Anleitung auf die Einhaltung der kurzen Atempause und warten Sie den Moment ab, bis der Atem von allein wieder einströmt.
- **Brillenträger** sollten die Brille bei dieser Übung abnehmen.
- Die sich anschließende wichtige Dunkelheit entspannt die Muskulatur wohltuend.

## Mentale Einstimmung

„Warst du schon mal in einem Glockenturm und hast beobachten können, wie der Glockenschwengel eine riesige Kirchturmglocke zum Klingen gebracht hat? Bei unserer heutigen Übung schwingen deine Augen wie ein großes Pendel hin und her. Damit deine Augen sich dabei wohlfühlen, schwingen sie ganz langsam von einer Seite zur anderen. Wenn du willst, kannst du dir dabei den Klang der Glocke vorstellen – ding ...... dong ...... ding ...... dong."

## Durchführung

„Setze dich entspannt auf deinen Stuhl ...... Deine Füße sind gut geerdet ...... Die Arme hängen locker nach unten ...... Deine Augen sind geschlossen ...... Bewege deine Augen langsam mit dem nächsten Einatmen von links nach rechts ...... Mit dem Ausatmen von rechts nach links ...... Kurze Atempause ...... Mit dem Einatmen von links nach rechts ....... Mit dem Ausatmen von rechts nach links ...... Kurze Atempause ...... Wechsle nun die Richtung! Bewege deine Augen mit dem Einatmen von rechts nach links ...... Mit dem Ausatmen von links nach rechts ...... Kurze Atempause ...... Mit dem Einatmen von rechts nach links ...... Mit dem Ausatmen von rechts nach links ...... Kurze Atempause ...... Lege dein Gesicht zum Schluss in deine Handflächen und entspanne deine geschlossenen Augen darin ...... Atme dabei in deinem Rhythmus weiter."

## Rücknahme

„Öffne deine Augen langsam und lasse sie sich wieder an die Helligkeit im Raum gewöhnen ...... Strecke und dehne dich so gut du kannst ...... Gähne noch mal tief ...... Schüttel deinen Körper zum Abschluss gut aus."

## Abschluss

„Sind deine Augen wieder frisch für den Unterricht?"

# Die Zitronenpresse

## Ziel der Übung

- **Aktiviert die Arm- und Handmuskeln,** funktionell alternativ auch mit den Zehen.

## Das sollten Sie wissen

- **Maximale Anspannung bringt maximale Entspannung.** Lassen Sie die Kinder richtig mit den Muskeln spielen, das macht Spaß.
- Die Einhaltung der kurzen Atempause verstärkt das Entspannungsgefühl.
- **Ruhige Musik** und Zitronenduft unterstützten die Entspannungsbereitschaft.
- Führen Sie diese Übung evtl. mit einer **echten Zitrone** vor.

## Mentale Einstimmung

„Hast du schon mal einen Zitronenkuchen gegessen? Dafür mussten einige Zitronen ausgepresst werden. Heute kannst du testen, wie viel Saft du aus deiner Zitrone pressen kannst. Deine Hände können stark sein. Reicht es für einen Zitronenkuchen?"

## Durchführung

„Setze dich entspannt auf deinen Stuhl ...... Beide Füße stehen fest auf dem Boden ...... Beide Arme hängen locker nach unten ...... Atme tief ein und mache eine starke Faust ...... Lockere die Hände und atme dabei aus ...... Kurze Atempause ...... Atme wieder tief ein und bilde eine Faust ...... Lockere sie mit dem Ausatem ....... Kurze Atempause ...... Presse noch ein paar Zitronen aus, damit der Kuchen saftig wird."

## Rücknahme

„Schüttel deine Hände kräftig aus und strecke dich weit in alle Richtungen."

## Abschluss

„Wie viele Zitronen hast du ausgepresst? Wie fühlen sich deine Hände und Arme an?"

# Der Körperbuzzer

## Ziel der Übung

- Diese Übung beinhaltet Elemente der Bewegung, **Konzentration sowie Körperwahrnehmung** und vermittelt den Kindern, wie man den Atem mit der eigenen Vorstellungskraft durch den ganzen Körper lenken kann.

## Das sollten Sie wissen

- Erklären Sie vor der Übung das Wort „**Buzzer**". Viele Kinder kennen ihn auch von dem Spiel „Halli Galli".
- Durch die Berührung lernen die Schüler, die Körperregionen bewusst zu spüren. Mit dem gelenkten Ausatmen lernen sie, bewusst loszulassen.
- Erweitern Sie bei Bedarf die Übung um **weitere Körperregionen**.
- Zaubern Sie vorher die Kinder ruhig, z. B. wer spricht, wird **versteinert**.
- Diese Übung können Sie in **jeder Position durchführen**, liegend, sitzend oder stehend.

## Mentale Einstimmung

„In Quiz-Sendungen kann man ihn häufig entdecken, den Buzzer. Weiß man die Antwort, drückt man ganz schnell drauf. Hast du einen Buzzer schon mal benutzt? Heute ist dein Körper der Buzzer. Du kannst zeigen, wie blitzschnell du sein kannst. Achte nur auf die Körperteile, die du gleich hören wirst, und berühre sie blitzschnell. Lenke anschließend den Atem beim Ausatmen genau in diese Stelle, das entspannt deinen Körper."

## Durchführung

„Stelle dich entspannt auf deinen Platz ...... Um dich herum ist viel Platz zum Nebenmann ...... Atme langsam tief ein und wieder aus ...... Los geht's ...... Berühre beide Knie ...... Atme tief dahin aus ...... Berühre deinen Hals ...... Atme tief dahin aus ...... Berühre deinen Rücken ...... Atme tief dahin aus ...... Berühre deinen rechten Fuß ...... Atme tief dahin aus ...... Berühre deine linke Schulter ...... Atme tief dahin aus ...... Berühre deinen linken Fuß ...... Atme tief dahin aus ...... Berühre deine rechte Schulter ...... Atme tief dahin aus ...... Berühre deinen Kopf ...... Atme tief dahin aus."

## Rücknahme

„Schüttel dich kräftig aus und setze dich wieder auf deinen Platz."

## Abschluss

„Konntest du deinen Atem beim Ausatmen immer an die angegebene Körperstelle lenken? Fühlst du die Entspannung im Körper?"

# Das Schmetterlingsblinzeln

## Ziel der Übung

- Kinder reiben sich bei **Überanstrengung oder Ermüdung** oft die Augen oder haben den sogenannten „starren Blick". Diese Übung sorgt dafür, dass sich der wichtige Tränenfilm über die Augen legt und diese wohltuend entspannt

## Das sollten Sie wissen

- Führen Sie diese Übung bei **Müdigkeitsanzeichen oder brennenden Augen** durch und lassen Sie die Schüler anschließend ein wenig in der „Kopfstützhaltung" ausruhen.
- Achten Sie darauf, dass die Schüler nur leicht blinzeln. Das feste Zusammenkneifen der Augen ist zu anstrengend und würde die jungen „Schüleraugen" überanstrengen.

## Mentale Einstimmung

„Bestimmt hast du schon mal einen Schmetterling beobachtet, wie er beim Nektarsaugen auf der Stelle die Flügel bewegt, ohne sich von der Stelle zu bewegen. Dabei flattern oder schwirren die Schmetterlinge sehr schnell.
Willst du mal für einen kurzen Augenblick ein Schmetterling sein? Wie sich ein etwas langsamerer Flügelschlag bei dir anfühlt, kannst du gut mit deinen Augenlidern und deinem Atem ausprobieren."

## Durchführung

„Setze dich entspannt auf deinen Stuhl und lasse alle Körperteile an dir hängen ...... Deine Muskeln sind sehr locker ...... Schüttel sie noch mal aus ...... ...... Richte deine Augen in die Ferne ...... Deine Augenlider sind nun deine Schmetterlingsflügel ...... Atme tief und langsam ein und blinzle ganz leicht und so schnell du kannst ...... Mit dem Ausatmen schließt du deine Augen ...... Mit dem Einatmen fängst du wieder zu blinzeln an ...... Lass deine Augenlider noch drei Atemzüge lang auf und ab ‚flattern'."

## Rücknahme

„Stütze deine Ellenbogen auf ...... Lege deinen Kopf in die Hände und ruhe dich mit abgedeckten Augen eine Minute aus. Rekele dich und strecke dich ganz weit nach oben und zur Seite."

## Abschluss

„Konntest du deine Augenlider auch so schnell bewegen wie ein Schmetterling die Flügel? Sind deine Augen gut entspannt?"

# Der Wackeldackel

## Ziel der Übung

- **Leichte und geschmeidige Bewegungen** lockern die – durch die meist statische Haltung im Unterricht verursachten – Spannungen im Schulter- und Nackenbereich.

## Das sollten Sie wissen

- Achten Sie auf eine **langsame und vorsichtige Ausführung** der Übung.

## Mentale Einstimmung

„Früher konnte man auf den Hutablagen mancher Autos kleine Tiere sehen, die beim Fahren mit dem Kopf nickten. Kennst du so einen ‚Wackeldackel'? Heute bist du ein kleiner Wackelhund, der ganz freundlich seine Mitmenschen grüßt und mit leichten Auf- und Abwärtsbewegungen seine Hals- und Nackenmuskeln lockert."

## Durchführung

„Setze dich aufrecht auf deinen Stuhl ...... Deine Arme hängen nach unten ...... Deine Fußsohlen stehen fest auf dem Boden ...... Richte deinen Blick waagerecht über deine rechte Schulter ...... Atme tief ein ...... Bewege deinen Kopf mit dem Ausatmen, langsam auf und ab wackelnd, nach links ...... Atme wieder ein ...... Bewege deinen Kopf, langsam auf und ab wackelnd, zur rechten Seite zurück ...... Atme wieder ein ...... Gehe wackelnd nach links ...... Hole tief Luft ...... Wackle nach rechts zurück ...... Wiederhole das Wackeln drei Atemzüge lang."

## Rücknahme

„Strecke und dehne dich ...... Streiche deinen Nacken mit einer Handmassage aus."

## Abschluss

„Hast du das Knirschen im Nacken gespürt? Wie fühlt er sich jetzt an?"

# Der Muskeldetektiv

## Ziel der Übung

- Diese Übung hilft, vorhandene **Anspannungen bewusster wahrzunehmen** und diese wieder zu lockern.

## Das sollten Sie wissen

- Gehen Sie vorher die **einzelnen Muskelgruppen** mit Ihren Schülern durch.
- Oder führen Sie zur Motivationssteigerung vor Beginn der Übung ein **kleines Muskel-Quiz** durch, z. B. „Wo sitzt der kleinste, der größte, der kräftigste, der längste Muskel?"
- Die Zählpausen von 1 bis 3 verdeutlichen die Anspannungszeit und sollten daher langsam und deutlich als Orientierungshilfe vorgesprochen werden.

## Mentale Einstimmung

„Jeder Mensch hat über 650 Muskeln. Den kleinsten Muskel findest du im Ohr. Der flächenmäßig größte Muskel ist der Pobackenmuskel, der kräftigste Muskel ist der Kaumuskel. Mehr als 50 Muskeln findest du allein in deinem Gesicht. Bist du wütend und schaust du grimmig, spannst du mehr Muskeln an als bei einem freundlichen Lächeln. Teste doch mal deine Muskeln. Wo kannst du deine Muskeln am deutlichsten spüren? Bist du bereit?"

## Durchführung

„Setze dich entspannt auf deinen Stuhl ...... Schließe deine Augen ...... Atme entspannt ein und aus ...... Spanne deinen Kaumuskel an ...... Beiße die Zähne fest zusammen ...... 1 – 2 – 3 ...... Lockere die Kaumuskeln ...... Atme entspannt weiter ...... Spanne die Fäuste an ...... 1 – 2 – 3 ...... Lockere die Fäuste ...... Atme entspannt weiter ...... Zwicke deine Pobackenmuskeln fest zusammen ...... 1 – 2 – 3 ...... Lockere deine Pobackenmuskeln ...... Atme entspannt weiter ...... Spanne deine Bauchmuskeln an ...... Mache deinen Bauch ganz groß ...... 1 – 2 – 3 ...... Lockere deinen Bauch ...... Atme entspannt weiter ...... Spanne alle Muskeln im Körper an ...... 1 – 2 – 3 ...... Lockere alle Muskeln ...... Atme entspannt eine paar Atemzüge weiter ......"

## Rücknahme

„Öffne deine Augen ...... Spüre deinen Körper ...... Strecke dich und schüttel alle deine Muskeln noch mal kräftig aus."

## Abschluss

„Welche Muskeln konntest du in deinem Körper am deutlichsten spüren?"

# Das Körpermemory

## Ziel der Übung

- Diese Übung schafft **Ruhe** und **erhöht die Aufmerksamkeit** der Kinder.

## Das sollten Sie wissen

- Während Sie ein Kleidungsstück vorgeben, überlegen die Kinder, zu welchem Körperteil es gehört, legen die Hände darauf, atmen tief ein und aus und spüren die Wärme.
- Führen Sie die Übung ganz **langsam und geräuschlos** durch, damit die Kinder die jeweiligen Körperstellen auch **achtsam wahrnehmen und spüren** können.
- Wenn möglich, dann sollten immer **beide Hände** zum Einsatz kommen.
- **Delegieren Sie** auch mal die Rolle des Anleiters an unruhige Kinder, das erhöht deren Aufmerksamkeit.
- Diese Übung kann **sitzend, liegend oder stehend** durchgeführt werden. Wechseln Sie öfter mal die Positionen und die Körperteile.

## Mentale Einstimmung

„Wann hast du das letzte Mal Memory gespielt? Beim Memory suchst du immer zwei Karten mit dem gleichen Bild, stimmt's? Bei unserem Körpermemory ist das ein klein wenig anders. Du berührst mit beiden Händen das zum Kleidungsstück passende Körperteil, atmest ganz tief ein und wieder aus und spürst dabei die Wärme, die von deiner Hand in diese Körperstelle einströmt."

## Durchführung

„Stelle dich gut geerdet auf deinen Platz ...... Lasse Arme und Hände locker nach unten hängen ...... Schließe deine Augen oder suche einen Punkt vor dir an der Wand ...... Atme tief ein und aus ...... Achte jetzt auf die Ansagen ......
**‚Mütze'** (Schüler berühren den Kopf, atmen tief ein, aus, spüren die Handwärme) ......
**‚Socken'** (S. berühren die Füße, atmen tief ein, aus, spüren die Handwärme) ......

**‚Schal‘** (S. berühren Hals und Nacken, atmen tief ein, aus, spüren die Handwärme) ......
**‚Hose‘** (S. berühren die Beine, atmen tief ein, aus, spüren die Handwärme).“

## Rücknahme

„Schüttel deine Arme und Beine kurz aus, strecke und dehne dich ausgiebig und spüre noch mal die Körperstellen, die du berührt hast.“

## Abschluss

„Gibt es Körperteile, die auch noch berührt werden wollen? Wie fühlt sich dein Körper jetzt an?“

# Die Zielscheibe

## Ziel der Übung

- Das abschließende **Palmieren** (Hände warm reiben und Handflächen auf die Augen legen) **entspannt die Augenmuskulatur** blitzschnell.

## Das sollten Sie wissen

- Unsere Augen werden durch den vermehrten Gebrauch von Medien mehr im Nahbereich gefordert. Dies hat zur Folge, dass unsere Augenmuskeln zunehmend verkürzen, was laut Aussagen von Augenärzten zu einer **Zunahme der Kurzsichtigkeit** bei Kindern führt. Das **Akkomodieren**, der Wechsel von Nah- und Ferneinstellung der Augen, ist eine **effektive Lockerungsübung** für die Augenmuskeln.

## Mentale Einstimmung

„Bestimmt hast du schon mal eine Zielscheibe gesehen, z. B. beim Darten oder beim Bogenschießen. Vielleicht hast du auch schon mal selbst einen Bogen oder einen Dartpfeil auf eine Scheibe gerichtet? Wir fangen heute mit einer ganz kleinen Zielscheibe an, die du immer bei dir hast: deine Handinnenfläche. Male dir zunächst, wenn du willst, in beide Handinnenflächen ein kleines Kreuz, das dir die Mitte deiner Zielscheibe anzeigt."

## Durchführung

„Setze oder stelle dich aufrecht hin ...... Halte mit der linken Hand dein linkes Auge zu ...... Strecke deinen rechten Arm weit nach vorn ...... schaue konzentriert dein kleines Kreuz auf der Handinnenfläche an ...... Atme langsam ein ...... Hole die Zielscheibe dabei langsam zu deinem rechten Auge ...... Atme aus und schließe deine Augen ...... Entspanne deine Augen einen Atemzug lang ...... Halte mit der rechten Hand dein rechtes Auge zu ...... Strecke deinen linken Arm weit nach vorn ...... schaue konzentriert dein kleines Kreuz auf der Handinnenfläche an ...... Atme langsam ein ...... Hole die Zielscheibe dabei langsam zu deinem linken Auge ......

Atme aus und schließe deine Augen....... Entspanne deine Augen einen Atemzug lang ...... Wiederhole die Übung auf beiden Seiten ein zweites/drittes Mal ...... Reibe deine Handflächen warm ...... Lege sie über deine Augen."

## Rücknahme

„Beende die Übung mit ausgiebigem Dehnen und Gähnen ...... Öffne anschließend deine Augen ...... Aktiviere deine Augenlidmuskeln mit leichtem Blinzeln."

## Abschluss

„Mit welchem Auge konntest du besser sehen/zielen?"

# Die Zielscheibe

© Eva Spanjardt

# Die freundlichen Minuten

## Ziel der Übung

- **Freundliche Worte** haben einen sehr positiven Effekt auf unser Verhalten und auf unsere Gefühle. Schenken Sie den Schülern häufig ein **wohlwollendes Lächeln** und spüren Sie, wie dankbar unsere kleinen und großen Mitmenschen Ihre Freundlichkeit und Achtsamkeit aufnehmen.

## Das sollten Sie wissen

- Bilden Sie zwei Schülerreihen und lassen Sie die Reihen aneinander vorbeigehen. Die Schüler kennen bestimmt **das Begrüßungsritual** bei Fußballspielen. Bei dieser Übung verschenken beide Reihen ein Lächeln.
- Achten Sie auf ein **stilles, lautloses Lächeln** mit geschlossenem Mund, damit die Übung nicht aus dem Ruder läuft.

## Mentale Einstimmung

„Hast du schon mal mitgezählt, wie oft du am Tag gelacht hast? Kinder lachen sehr viel häufiger als Erwachsene. Wie oft lachen denn die Erwachsenen in deinem Umfeld – deine Eltern, deine Lehrer, deine Oma? Wann spannst du mehr Gesichtsmuskeln an, wenn du lachst oder wenn du grimmig schaust? – Richtig, das Lachen entspannt die Gesichtsmuskeln und bildet in deinem Körper einen Stoff, der dich glücklich und zufrieden macht. Man nennt den Stoff auch Glückshormon. Aber nicht nur dich selbst, auch deine Mitmenschen kannst du mit einem freundlichen Lächeln glücklich machen. Probiere es doch mal aus und verschenke heute dein schönstes Lächeln. Mit jedem tiefen Atemzug zauberst du dir ein Lächeln auf deine Lippen und mit jedem Ausatmen verschenkst du es an einen Mitschüler. Schaue deinen Mitschülern dabei in die Augen."

## Durchführung

„Stelle dich bequem in deine Reihe ...... Atme tief ein ...... Zaubere ein Lächeln auf deine Lippen ...... Suche Blickkontakt zu einem Mitschüler in der gegenüberliegenden Reihe ...... Atme aus und verschenke dein schönstes Lächeln ...... Atme wieder tief ein ...... Zaubere ein Lächeln auf deine Lippen ...... Gehe weiter zum nächsten Mitschüler ...... Atme aus und schenke diesem Mitschüler dein Lächeln ...... Gehe die Reihe bis zum Ende durch und lächle so gut du heute kannst."

## Rücknahme

„Schüttel zum Abschluss lächelnd deinen Körper locker."

## Abschluss

„Wie hat sich die Freundlichkeit deiner Mitschüler angefühlt? Fühlst du dich nun gut?"

# Die Sonnenatmung

## Ziel der Übung

- Die Übung **lockert die Rumpfmuskulatur** und unterstützt die Bauchatmung.

## Das sollten Sie wissen

- **Achten Sie auf genügend Abstand** zum Nebenmann.
- Eine **langsame und bewusste Durchführung** ist wichtig, damit Ihre Schüler ihre Muskeln wahrnehmen und den langsamen Atemrhythmus einhalten können.

## Mentale Einstimmung

„Hast du schon mal einen Sonnenaufgang beobachtet? Ganz langsam taucht im Osten die Sonne am Morgen am Himmel auf, erreicht mittags den Höchststand und geht gegen Abend im Westen wieder langsam unter. Heute bist du unsere Sonne."

## Durchführung

„Begib dich in die Hockstellung ...... Mache dich ganz klein ...... Lasse nun die Sonne aufgehen ...... Atme tief ein ...... Breite deine Sonnenstrahlen-Arme zur Seite aus ...... Richte deinen ganzen Körper langsam auf ...... strecke deine Arme weit nach oben ...... Atme tief aus und wieder ein ...... Lasse die Sonne mit dem Ausatem langsam untergehen ...... Komme in die Hockstellung zurück ...... Wiederhole den Sonnenlauf noch ein paar Atemzüge lang."

## Rücknahme

„Beende die Übung mit einem kräftigen Ausschütteln deines Körpers."

## Abschluss

„Fühlt sich dein Körper nun wieder kraftvoll wie die Sonne an?"

# Die Sonnenatmung

# Das magische Auge

## Ziel der Übung

- **Regelmäßig durchgeführt,** stärkt und trainiert diese Übung wohltuend die Augenmuskulatur.

## Das sollten Sie wissen

- Wem diese Übung nicht sofort gelingt, kann versuchen, die ausgestreckten Daumen langsam näher an die Augen heranzuführen, bis sie fast die Nasenspitze berühren. Dann werden die Daumen wieder langsam ca. 20 cm zurückbewegt. Wichtig ist, dass durch die Daumen hindurchgeträumt wird.
- Stellen Sie die Übung als **Experiment** vor. Mal gelingt ein Experiment, mal eben nicht.
- **Auch mit Brille oder Kontaktlinsen** kann man diese Übung gut durchführen.

## Mentale Einstimmung

„Es gibt interessante Zaubertricks, die wir mit viel Üben erlernen können. Diesen Trick kann aber jeder bereits nach ein paar Sekunden schaffen. Willst du mal versuchen, dir einen dritten Daumen zu zaubern? Wer schafft diesen Trick? Probiere es aus!"

## Durchführung

„Setze dich gerade auf deinen Stuhl ...... Deine Füße sind gut geerdet ...... Deine Arme sind nach oben angewinkelt ...... Die Ellenbogen und Unterarme berühren sich ...... Balle die Hände zu Fäusten ...... Deine Handballen berühren sich ...... Deine Daumen sind nach oben gestreckt ...... Deine Daumennägel sind auf Augenhöhe ...... Richte deinen Blick ein paar Atemzüge lang ruhig und gedankenversunken auf beide Daumen ...... Schaue entspannt durch beide Daumen auf einen dahinterliegenden Punkt, z. B. auf die Tafel oder an die Wand ...... Entdecke neben deinen zwei Daumen noch einen dritten Daumen ...... Stelle deine Augen wieder scharf und atme entspannt weiter ...... Wiederhole die Übung noch ein paar Atemzüge lang ......"

## Rücknahme

„Massiere zum Abschluss mit beiden Händen leicht den Augenbereich und entspanne deine geschlossenen Augen für einen kleinen Moment, indem du sie mit deinen warm geriebenen Händen wärmst."

## Abschluss

„Konntest du den dritten Daumen erkennen?"

# Das magische Auge

© Eva Spanjardt

# Das Körpersudoku

## Ziel der Übung

- Ein **überkreuztes Berühren der Körperteile** fördert die Aufmerksamkeit und aktiviert beide Gehirnhälften. Durch die Mobilisierung der Muskeln eignet sich die Übung auch gut zur **Auflockerung zwischen den Stunden**.

## Das sollten Sie wissen

- Viele Kinder haben bestimmt schon mal ein Sudoku gelöst. **Erklären Sie** trotzdem kurz **den Begiff „Sudoku"**.
- Durch eine detailliertere Einteilung der Körperregionen oder die Nennung von Zahlen statt Körperteilen lässt sich die **Schwierigkeitsstufe erhöhen**.
- Die Bildung von Sudoku-Teams erhöht die **Mitmachbereitschaft** und das Selbstwertgefühl von schwächeren und schüchternen Kindern, da der Fokus auf die Teamleistung gelenkt wird.

## Mentale Einstimmung

„Hast du schon mal ein Sudoku gelöst, vielleicht ein Rechtschreib-Sudoku oder ein Rechen-Sudoku? Heute kannst du an einer neuen Art Sudoku testen, wie gut du dich konzentrieren kannst und wie angenehm beruhigend so ein Sudoku sein kann.
Bei unserem Körper-Sudoku teilst du deinen Körper in sechs Bereiche ein:
**1 = Füße, 2 = Knie, 3 = Becken, 4 = Bauch, 5 = Schultern, 6 = Kopf.**

Fünf Bereiche des Körpers werden langsam und deutlich vorgegeben. Du berührst dabei diese Körperteile mit überkreuzten Armen und überlegst, welcher Bereich von den sechs Bereichen noch fehlt. Wie lange das Sudoku dauert, könnt ihr selbst bestimmen."

## Durchführung

„Stelle dich aufrecht mit ausreichend Abstand zum Nachbarn an deinen Platz ...... Berühre mit überkreuzten Händen deinen Kopf ...... deine Knie ...... deine Schultern ...... deinen Bauch ...... dein Becken ...... Welcher Bereich fehlt?

Berühre auch diesen blitzschnell über Kreuz ...... Schüttle dich aus ...... Atme durch ...... Konzentriere dich auf die nächste Ansage ......"

## Rücknahme

„Schüttle dich zum Abschluss noch mal kräftig aus ...... Dehne und strecke dich in alle Richtungen."

## Abschluss

„Konntest du den fehlenden Bereich immer sofort finden? Wie fühlt sich dein Körper an?"

# Das Körpersudoku

© Eva Spanjardt

# Imaginative Übungen

# Einführung

Bei einer imaginativen Übung – z. B. einer Fantasiereise – werden die Gedanken von außen gesteuert, der Geist bleibt dabei hellwach, die Muskeln werden locker, die Herzschlagrate, die Atem- und Pulsfrequenz nehmen ab, der ganze Körper entspannt.

Kinder können sich beonders gut in Geschichten wiederfinden und sich dabei vollkommen fallen lassen. Eine Fantasiereise, die langsam mit angenehmer, leiser Stimme vorgelesen wird, lenkt die Gedanken der Schüler schnell in ihre Traumwelt. Durch die Bilder, die vor dem inneren Auge entstehen, lernen die Kinder die Gefühle, die z. B. vor aufregenden Ereignissen und bevorstehenden Aufgaben/Klassenarbeiten entstehen können, wahrzunehmen, besser zu verstehen und zu verarbeiten.

Die bewusst kurz gehaltenen imaginativen Entspannungsgeschichten geben den Kindern die Möglichkeit, ihre Sinne zu aktivieren und das eigene Wohlbefinden blitzschnell zu steigern.

Besonders im Klassenverband gelingt es durch einen regelmäßigen und dauerhaften Einsatz entspannungsfördernder Fantasiereisen, die den Schulalltag oft begleitenden körperlichen und seelischen Probleme loszulassen, die Aufmerksamkeit auf sich zu lenken und die oft verschlossenen Türen im Kopf wieder zu öffnen. Kinder lernen auch rasch, wie sie sich mithilfe ihrer Fantasie die Welt und das eigene Erleben angenehmer und zufriedener „denken" können.

# Der Fantasieaufzug

## Ziel der Übung

- Diese Übung entspannt **nach langen Konzentrationsphasen.**

## Das sollten Sie wissen

- Geben Sie den Kindern etwas **Zeit, ihre Erlebnisse mitzuteilen**, das entspannt zusätzlich.
- Bei dieser Übung wäre das **Einspielen leiser Entspannungsmusik** ratsam. Sie unterstützt die Vorstellungskraft und steigert die Entspannungsbereitschaft Ihrer Schüler.

## Mentale Einstimmung

„Bist du schon mal mit einem Aufzug gefahren? Bestimmt hast du den Knopf drücken dürfen, der dich in ein anderes Stockwerk gebracht hat. Heute lade ich dich ein, mit einem besonderen Aufzug zu fahren. In unserem Fantasieaufzug findest du statt Zahlenknöpfe, große Tasten, auf denen steht, welches Erlebnis auf dich wartet. Höre genau zu und drücke dann gedanklich die Taste für das Stockwerk, in dem du aussteigen willst."

## Durchführung

„Stockwerk 1 – **die Spielplatz-Etage** ...... Hier findest du den schönsten Spielplatz, den du je gesehen hast ......

Stockwerk 2 – **die Massage-Etage** ...... Hier wirst du wohltuend massiert und kannst spüren, wie sich dein ganzer Körper entspannt ......

Stockwerk 3 – **der fliegende Teppich** ...... Hier schwebst du ganz gemütlich über deinen Wohnort und kannst alles von oben beobachten ......

Stockwerk 4 – **die Freunde-Etage** ...... Hier triffst du alle Freunde wieder, die du schon lange nicht mehr gesehen hast ......

Entscheide dich nun für ein Stockwerk und sei gespannt, was dich dann in deiner Fantasie erwartet ......“ *Einspielen der Musik*

## Rücknahme

„Verabschiede dich nun wieder von deinem Stockwerk und komme von deinem Aufzug direkt in unser Zimmer zurück ...... Strecke deinen Körper und schüttel dich kurz aus.“

## Abschluss

„Welches Stockwerk hast du dir ausgesucht? Was hast du erlebt?“

# Die Atemschaukel

## Ziel der Übung

- Diese Übung entspannt und sorgt für ein gutes körperliches Wohlbefinden durch eine **bewusste Atmung**.

## Das sollten Sie wissen

- Bei einem **hohen Anteil an unruhigen Kindern** empfehle ich, die Übung im Liegen zu machen. Die großflächige Erdung des gesamten Körpers führt zu einer schnelleren Entspannungsbereitschaft.
- Sie können die Übung auch im Stehen gut durchführen. **Eine intensive Erdung** beider Füße ist jedoch wichtig, um einem Umfallen vorzubeugen.
- Passen Sie die **Länge der Sprechpausen** der Entspannungsfähigkeit Ihrer Klasse an. Sie können die Kinder auch gerne länger „schwingen" lassen.
- Führen Sie Atemübungen stets bei **geöffnetem Fenster** oder nach guter Belüftung durch.

## Mentale Einstimmung

„Du hast mit deiner Klasse an einem Wettwerb teilgenommen. Ihr habt den ersten Preis gewonnen: Eine riesig große Gemeinschaftsschaukel, die Platz für alle Kinder bietet. Das Besondere an dieser Schaukel ist, dass sie nur mit eurer Atemluft bewegt werden kann."

## Durchführung

„Schließe deine Augen ...... Stelle dir vor, du sitzt zusammen mit deinen Mitschülern auf eurer Atemschaukel ...... Zusammen atmet ihr nun ganz langsam und tief in den Bauch ein ...... Die Schaukel fängt an, nach hinten zu schwingen ...... Mit jedem langen Ausatem schwingt die Schaukel nach vorn ...... Dein Bauch wird wieder dünn ...... Du schwingst vor und zurück ...... Die Schaukel ist groß ...... Atme ganz langsam ...... Fülle deinen Bauch mit Atemluft ...... Du schließt die Augen und spürst die

Luft im Gesicht ...... Mit jedem Atemzug schaukelst du höher und höher ...... Fühle dich frei wie ein Vogel ...... Jetzt könnt ihr die Schaukel langsam ausschwingen lassen ...... Der Atem wird langsamer und langsamer ...... Die Schaukel bleibt stehen."

## Rücknahme

„Steige gedanklich aus der Schaukel ...... Öffne deine Augen ...... Strecke dich aus ...... Schüttel Arme und Beine aus und spüre einen Augenblick die Schwere deines Körpers."

## Abschluss

„Konntest du die Schaukel mit deinem Atem bewegen?"

# Das Zauberzelt

## Ziel der Übung

- Diese Übung eignet sich nach großen Anstrengungen zum **körperlichen und mentalen Loslassen**.

## Das sollten Sie wissen

- Geben Sie den Kindern im Anschluss **Zeit, kurz zu berichten**.
- **Leise Entspannungsmusik** unterstützt das „Loslassen".
- Das Ausklingen der Musik zeigt den Kindern das Ende der Reise an.
- Das **Ablegen des Oberkörpers auf dem Tisch** erleichtert manchen Kindern das schnelle Entspannen.

## Mentale Einstimmung

„Heute hat deine Klasse einen Wandertag. Bei sonnigem Wetter spaziert ihr zu einem nahe gelegenen Zauberzelt. Du hast schon viel über dieses Zauberzelt gehört und freust dich sehr, es heute besichtigen zu dürfen. Was du in diesem Zelt erleben kannst, erfährst du, sobald es in deiner Klasse ganz still geworden ist."

## Durchführung

„Schließe deine Augen ...... Lasse dich entspannt in deine Sitzlehne fallen ...... Atme tief ein ...... Und wieder aus ...... Noch mal ein und wieder aus ...... Deine Gedanken machen jetzt einen Ausflug ...... Stelle dir ein wunderschönes, buntes Zelt vor ...... mit blauen, grünen, roten und orangenen Farben ...... Du kommst an eine große, goldene Tür ...... Die Tür öffnet automatisch ...... Im Inneren entdeckst du einen langen Gang mit vielen Spiegeln ...... Du gehst zu einer Tür, die mit roten Herzen und roten Rosen verziert ist ...... Auf der Tür stehen Wörter ...... Du kannst sie lesen ...... ‚Liebe und Geborgenheit'...... Neugierig öffnest du die Tür ...... Du bist sehr überrascht ...... Schwebende Wattewolken fliegen kreuz und quer durch das Zelt ...... Du suchst dir eine Wolke aus ...... Sie trägt dich ganz sicher durch das Zelt ......

Du kuschelst dich tief in die weiche Watte ...... Du fühlst dich wohl und genießt es, auf der Wolke zu schweben ...... Viele Kinder wollen auf dieser Wolke schweben ...... Komme wieder auf den Boden zurück und steige dankbar und gut gelaunt von deiner Wolke ......"

## Rücknahme

„Verabschiede dich wieder von deinem wunderbaren Zauberzelt ...... Öffne deine Augen ...... Komme mit deinen Gedanken hier in unseren Raum zurück ...... Strecke und dehne deinen Körper so gut du kannst."

## Abschluss

„Hat es dir in dem Zauberzelt gefallen?"

# Die Pauseninsel

## Ziel der Übung

- Diese Ruhereise entspannt die Kinder, z. B. **vor Klassenarbeiten**, und kann auch sitzend durchgeführt werden.

## Das sollten Sie wissen

- Gestalten Sie die **Sprechpausen etwas länger**, damit die Kinder ihre Fantasie frei entfalten können.

## Mentale Einstimmung

„Hast du Lust, dich mal zwischen den Stunden zu entspannen und wieder neue Kraft aufzutanken? Dann lade ich dich ein, mit mir auf eine Insel zu kommen, auf der es dir bestimmt gut gefallen wird."

## Durchführung

„Lege dich entspannt auf den Rücken ...... Schließe deine Augen ...... Seit heute gibt es auf deinem Schulhof eine neue Pauseninsel ...... Du kannst die Insel über eine Brücke erreichen ...... Es gibt leckeres Inselessen ...... leckere Inselgetränke ...... Ein Piratenschiff liegt mitten auf der Insel ...... Neugierig kletterst du auf das Schiff ...... Du kannst sehr viel darauf entdecken ...... Fühle dich wie im Urlaub ......"

## Rücknahme

„Die Pause ist leider zu Ende ...... Öffne deine Augen ...... Schüttel dich zum Abschluss kräftig aus ...... Dehne und strecke dich in alle Richtungen."

## Abschluss

„Hast du dich wohlgefühlt auf deiner Pauseninsel? Was hat dir besonders gut gefallen?"

# Das Steine-Flitschen

## Ziel der Übung

- Diese Ruhereise **entspannt** die Kinder, z.B. **vor Klassenarbeiten**, und kann auch sitzend durchgeführt werden.

## Das sollten Sie wissen

- **Leise Musikuntermalung** fördert die Ruhebereitschaft und die Konzentrationsfähigkeit.
- Erklären Sie den Schülern bei Bedarf den **Begriff „Flitschen"**. Flitschen steht für: Steine über das Wasser werfen, mit dem Ziel, sie so oft wie möglich aufkommen und weiterfliegen zu lassen.

## Mentale Einstimmung

„Wann warst du das letzte Mal an einem See? War dein See groß oder klein? Bestimmt bist du schon mal mit einem Boot gefahren oder in einem See geschwommen. Es gibt ganz viele Möglichkeiten, was man auf einem See alles machen kann. Was fällt dir alles ein? Heute möchte ich dir ein nettes Spiel zeigen, das du bestimmt auch schon mal ausprobiert hast."

## Durchführung

„Lege dich bequem auf deine Unterlage ...... Schließe deine Augen, wenn du kannst ...... Achte auf deine Atemzüge ...... Und komme in deinen Gedanken an einen wunderschönen See ...... Der Weg zum Ufer ist voll von Steinen ...... Du findest ganz viele flache Steine ...... Steckst sie in deine Hosentasche ......
Du gehst damit ans Ufer ...... Das Wasser ist klar und ganz ruhig ...... Du schmeißt die Steine ins Wasser ...... Es bilden sich Kreise ...... Du beobachtest die Kreise ...... Sie werden immer größer ...... Du nimmst einen flachen Stein ...... Du lässt ihn auf dem Wasser flitschen ...... 2-, 3-, 4-, 5-mal ...... Du versuchst es noch mal ...... diesmal

mit einem größeren Stein ...... Zähle mit, wie oft dein Stein aufkommt, bis er untergeht ...... Nur das Plätschern deiner Steine ist zu hören ...... Freue dich über deine Flitsch-Erfolge."

## Rücknahme

„Verabschiede dich von deinem See ...... Öffne deine Augen und komme in unseren Klassenraum zurück ...... Schüttel deine Arme aus, bewege deine Füße und Beine und komme in die Sitzposition zurück."

## Abschluss

„Hat das Steine-Flitschen deine Gedanken beruhigt?"

# Der Gedankenzauberer

## Ziel der Übung

- Schon Grundschüler setzen sich täglich mit negativen Gedanken unter Druck, fühlen sich dadurch schwach und mutlos. Mit dieser Übung lernen Ihre Schüler die Kraft positiver Gedanken kennen und wie es blitzschnell gelingen kann, ihr Wohlbefinden und Selbstbewusstsein zu stärken.

## Das sollten Sie wissen

- Sprechen Sie mit Ihrer Klasse vor dieser Übung über **gute Gefühle** und erklären Sie, dass der Körper durch angenehme Gedanken und Gefühle in der Lage ist, Glückshormone (Endorphine) im Körper auszuschütten.
- Bringen Sie Ihre Schüler so oft wie möglich durch **kleine Witze oder lustige Geschichten** zum Lachen. Das entspannt und befreit die Schüler von Anspannungen, die oftmals vor Klassenarbeiten zu beobachten sind.
- Geben Sie vor allem eher zurückgezogenen, ruhigeren oder emotional belasteten Kindern die **Möglichkeit, ihre positiven Erlebnisse zu erzählen**, das stärkt das Selbstwertgefühl.

## Mentale Einstimmung

„Wir Menschen denken täglich viele Tausend Gedanken. Mal sind sie gut, mal ängstigen sie uns. Doch wer bestimmt eigentlich, wie und was wir denken? Richtig! Wie wir denken und uns dabei fühlen wollen, kann jeder von uns meistens selbst beeinflussen. Bei unserer heutigen Übung lernst du, wie man gute Gedanken in seinen Kopf zaubern kann. Hast du Lust, es mal auszuprobieren?“

## Durchführung

„Setze oder lege dich entspannt auf deinen Platz und versuche, so gut du kannst still zu lächeln ...... Atme dein Lächeln ganz tief ein und erinnere dich so gut du kannst an ein schönes Erlebnis ...... Vielleicht an deinen Geburtstag? ...... Oder an deinen ersten Schultag ...... Oder an einen Ausflug mit der Klasse oder der Familie? ...... Lächelst du noch? ...... Atme das Lächeln noch mal ganz tief ein ...... Über was hast du dich in letzter Zeit sehr gefreut? ...... Vielleicht über ein schönes Wort? ...... Vielleicht über ein kleines Geschenk? ...... Vielleicht über eine Einladung? ...... Spürst du dein Lächeln noch auf den Lippen? ...... Atme es noch mal tief ein ...... Spüre, wie du dich gerade fühlst ...... Hast du gute Gedanken in deinen Kopf zaubern können?

## Rücknahme

„Strecke und dehne dich so gut du kannst ...... Öffne deine Augen und spüre, wie glücklich du dich fühlst."

## Abschluss

„Willst du kurz erzählen, wie du dich gefühlt hast, oder ein Glückbild malen?"

# Das neue Schulfach

## Ziel der Übung

- Die Kinder lernen, mentale Rückzugsorte als ein Wohlgefühl wahrzunehmen.

## Das sollten Sie wissen

- Finden Sie heraus, welche Räume Ihre Schüler als angenehm empfinden. Vielleicht kann das Klassenzimmer zum Lernraum mit „Glücksfaktor" umgestaltet werden.

## Mentale Einstimmung

„An deiner Schule gibt es ein neues Fach. Es heißt ‚Glück und Freude' und wird im Glückszimmer im Dachgeschoss unterrichtet. Du bist gespannt, was du lernen kannst!"

## Durchführung

„Lege oder setze dich bequem hin ...... Schließe deine Augen ...... Atme tief ein und wieder aus ...... Das Glückszimmer befindet sich im Dachgeschoss ...... Steige die vielen Treppenstufen hinauf ...... Du kommst an eine bunt glänzende Tür ...... Vorsichtig öffnest du die Tür ...... Eine sonnige Terrasse erwartet dich ...... Es riecht angenehm frisch nach Zitronen und frischer Luft ...... Die Terrasse steht voll mit Schaukeln ...... Suche dir eine aus und setze dich darauf ...... Beginne, zu schaukeln ...... Immer höher und höher ...... Genieße den fantastischen Ausblick ...... Spüre das Kribbeln in deinem Bauch ...... Fühle dich ganz wohl und glücklich ......"

## Rücknahme

„Lasse die Schaukel langsamer werden ...... Öffne deine Augen ...... Steige von deiner Schaukel ...... Dehne und strecke deine Schaukel-Arme und Schaukel-Beine."

## Abschluss

„Wie findest du das neue Schulfach?"

# Die Farbgedanken

## Ziel der Übung

- **Mentale Entspannung** und **Regeneration zwischen den Stunden**

## Das sollten Sie wissen

- **Grün, Dunkelblau und Gelb** eignen sich wunderbar zur **Entspannung**. Mithilfe von z. B. Farblampen kann eine Steigerung des Wohlbefindens erreicht werden.
- Sie können diese Übung auch sitzend mit abgelegtem Oberkörper durchführen.

## Mentale Einstimmung

„Hast du gewusst, dass Farben mitbestimmen, wie wir uns fühlen? Hast du eine Lieblingsfarbe? Willst du ausprobieren, wie Farben deine Stimmung beeinflussen?

## Durchführung

„Lege dich bequem auf deine Unterlage ...... Schließe deine Augen ...... Dein ganzer Körper wird schwerer und schwerer ...... Gehe mit deinen Gedanken zu deinem Kopf ...... Stelle dir die Farbe Grün so leuchtend wie möglich vor, wie das Grün eines Apfels ...... Atme das Grün tief ein ...... Spüre, wie dich das Grün beruhigt und entspannt ...... Stelle dir die Farbe Blau so leuchtend wie möglich vor, wie das Blau des Meeres ...... Atme das Blau tief ein ...... Spüre, wie dich das Blau zufrieden und glücklich werden lässt ...... Stelle dir die Farbe Gelb so leuchtend wie möglich vor, wie das Gelb einer Zitrone ...... Atme das Gelb tief ein ...... Spüre, wie dich das Gelb fröhlich werden lässt ......“

## Rücknahme

„Atme noch einmal tief ein ...... Öffne deine Augen ...... Schüttel Arme und Hände fest.“

## Abschluss

„Welche Farbe hast du dir am besten vorstellen können?“

# Die Weihnachtsdüfte

## Ziel der Übung

- Mit dieser Übung lenken Sie die Wahrnehmung auf angenehme und für Kinder positive Verknüpfungen mit dem Ziel, die **Konzentration zu steigern und ein körperliches Wohlfühlen** zu vermitteln.

## Das sollten Sie wissen

- Sind wir einem Duft länger und in gleichbleibender Intensität ausgesetzt, dann nehmen wir den Geruch immer schwächer bzw. gar nicht mehr wahr. **Lüften Sie das Klassenzimmer** deshalb ausgiebig und regelmäßig.
- **Konzentrationsfördernde und harmonisierende Düfte** von Zitrusfrüchten, Geranien, Lavendel und Vanille können die Lernatmosphäre positiv beeinflussen, inspirierend wirken und die Wahrnehmungsfähigkeit und das Wohlgefühl der Schüler fördern.
- Düfte sind **eng mit unseren Gefühlen verknüpft** und können sekundenschnell ein positives oder negatives Gefühl erzeugen.

## Mentale Einstimmung

„Es gibt unendlich viele Düfte, z. B. blumige Düfte, wie Rosen- und Lavendelduft, würzige Düfte, wie Zimt und Nelken. Es gibt fruchtige Düfte, wie Zitronen- und Mandarinenduft. Und es gibt orientalische Düfte, wie Weihrauch. Unser Gehirn hat mehrere Hundert Düfte gespeichert und kann sofort eine Verbindung zu einem Erlebnis oder einem festlichen Ereignis herstellen. Welche Düfte und Verbindungen bei dir im Kopf gespeichert sind, erfährst du bei dieser Übung."

## Durchführung

„Setze oder lege dich entspannt auf deinen Platz ...... Schließe deine Augen und öffne deine Nase ...... Atme tief ein und wieder aus ...... Heute besuchst du mit deiner Klasse den Weihnachtsmarkt ...... Von Weitem kannst du bereits die süßen, leckeren Düfte riechen ...... Der Duft süßer Mandeln und Lebkuchen strömt tief in deine Nase ...... Würziger Glühweinduft streift immer wieder an dir vorbei ...... Der Duft leckerer Bratwürste lässt dir das Wasser im Mund zusammenlaufen ...... Würziger Zimtduft lenkt deine Blicke zu einem Crêpes-Stand ...... Welche Gerüche kannst du noch erschnuppern?" ...... (1 Min.)

## Rücknahme

„Beende deinen Spaziergang durch den Weihnachtsmarkt ...... Atme die Düfte noch mal kräftig in deine Nase ...... Öffne deine Augen ...... Strecke und rekele dich in alle Richtungen ...... Erzähle von deinem Lieblingsduft."

## Abschluss

„Welche Düfte verbindest du mit Weihnachten? Konntest du dir alle Düfte vorstellen? Welche Düfte findest du am angenehmsten?"

# Der magische Luftballon

## Ziel der Übung

- Bei dieser Fantasiereise tauchen die Schüler blitzschnell in eine wohltuende Gedankenwelt ein und lernen dabei, wie schnell man seine **Gedanken loslassen** kann.

## Das sollten Sie wissen

- Gedanken und Gefühle lassen sich in Sekundenschnelle verändern. Allein wie man die **Gedanken bewertet**, entscheidet darüber, wie man sich anschließend fühlt. Machen Sie sich diese Tatsache zunutze, wenn es darum geht, die Stimmung in der Klasse positiv zu verändern.

## Mentale Einstimmung

„Bei vielen Hochzeitsfeiern oder Geburtstagen lässt man ganze Trauben von bunten, mit Helium gefüllten Luftballons in den Himmel steigen. Warst du auch schon mal dabei und hast ganz verzaubert den tänzelnden Ballons hinterhergesehen? Versuche, dich daran zu erinnern oder dir die vielen bunt leuchtenden Luftballons in deiner Fantasie vorzustellen."

## Durchführung

„Setze oder lege dich bequem auf eine Unterlage ...... Deine Muskeln sind ganz locker ...... Deine Augen sind geschlossen ...... Lasse in deiner Fantasie ganz viele Luftballons in den blauen Himmel aufsteigen ...... mit lustigen Formen ...... große und kleine Ballone ...... dicke und dünne ...... Sie fliegen steil nach oben ...... leuchten in ihren bunten Farben ...... rot ...... grün ...... gelb ...... blau ...... orange ...... Die langen Schnüre an den Ballons tanzen im Wind ...... Schaue ihnen nach! ...... Mit jedem Atemzug steigen sie weiter und weiter nach oben ...... Ganz verspielt lassen sie sich treiben ...... zaubern ein buntes Muster in den Himmel ...... werden dabei immer kleiner und kleiner ...... wie bunte Stecknadelköpfe ...... bis sie ganz verschwinden ...... sich mit der Farbe des Himmels vermischen ...... du sie nicht mehr erkennen kannst ......"

## Rücknahme

„Komme mit deinen Gedanken wieder in unseren Klassenraum zurück ...... Öffne langsam deine Augen ...... Bewege deine Arme und Beine ...... Schüttel dich gut aus."

## Abschluss

„Welche Formen und Farben hatten deine Luftballons?'

# Auf Achtsamkeit basierende Übungen

# Einführung

Achtsamkeit als gesteigerte Form von Aufmerksamkeit wird für unsere Kinder im Schulalltag immer wichtiger. Stetig wachsende Anforderungen im Unterricht und eine ständig zunehmende Unterhaltungsflut und Informationsschwemme verlangen von unseren Schülern ein Höchstmaß an Disziplin und Zurückhaltung.

War Achtsamkeit noch vor Jahren eher der esoterischen Richtung zugeordnet, erkennen immer mehr Ärzte und Therapeuten die Wirksamkeit einer auf Achtsamkeit basierenden Therapie.

Aber auch in Schulen und anderen Bildungseinrichtungen wächst die Zahl der Lehrer, die ein Achtsamkeitstraining bei Unterrichtsstörungen, ausgelöst durch vermehrt auftretenden Stress und wachsende psychische Störungen bei Schülern, durchweg als positiv und leistungsfördernd erkennen und einsetzen.

Achtsame, vom Atem getragene Entspannungsübungen ermöglichen den Schülern, frühzeitig den wachsenden Anforderungen in Unterricht und Freizeit gelassener und zufriedener zu begegnen, um somit Körper und Geist vor Überforderung und Dauerstress zu schützen.

Die Schüler lernen, ganz bewusst auf ein gesundes und zufriedenes Körpergefühl zu achten, ihre allgegenwärtigen Gedanken wahrzunehmen und diese bewertungsfrei wieder loszulassen.

Achtsames Hören, Riechen, Fühlen, Schmecken und Sehen lenkt dabei die Aufmerksamkeit der Schüler in die beabsichtigte und wohltuende Phase des erlaubten „Nichtstuns" und stärkt ein gesundheitsförderndes Verhalten.

Die achtsame Wahrnehmung fördert neben der Konzentration und der Aufmerksamkeit auch das Sozialverhalten der Schüler. Wer lernt, achtsam mit sich selbst zu sein, wird auch seinen Mitmenschen respektvoll begegnen.

Achtsam durchgeführte Entspannungspausen sind gesundheitsfördernd und leistungssteigernd, da sie durch die für unser Gehirn reinigende Funktion freie Kapazitäten für neues Wissen und neue Aufgaben schaffen.

# Das Ruheritual

## Ziel der Übung

- **Achtsamkeits-Rituale unterstützen die Entspannungsbereitschaft** und die **Konzentration** vieler Schüler.

## Das sollten Sie wissen

- Die folgenden zwei Entspannungsformeln können unruhigeren und entspannungsunerfahrenen Kindern den Zugang zur Entspannung erleichtern und eignen sich daher als kurzes Einstiegsritual für alle Übungen des Buches.

  **Entspannungsformel1:**
  „Ich lass die Muskeln einfach locker – sitz entspannt auf meinem Hocker.
  Atme ein und atme aus – fühl mich wie im Schneckenhaus.
  Schließ die Augen, wenn ich will – plötzlich ist es mäuschenstill."

  **Entspannungsformel 2:**
  „In meinem Kopf, da geht es zu – es plappern 1 000 Affen.
  Ich bring die Tierchen kurz zur Ruh – ich weiß, ich kann das schaffen.
  Ich atme still, das Denken ruht – ich spüre schon, wie gut das tut."

- Sie können die Ruhephase auch mit einer **Klangschale oder einem Gong** einleiten.
- Öffnen Sie während der Übung, wenn möglich, die Fenster, damit auch **Außengeräusche hörbar werden**.

## Mentale Einstimmung

„Hast du schon mal eine vollkommene Stille erlebt? Wenn wir glauben, es sei ganz ruhig, können wir bei genauerem Hinhören noch viele Geräusche hören. Höre doch mal in die Stille der Klasse hinein und merke dir mindestens fünf Geräusche, die du nur hören kannst, wenn es in deinem Kopf und um dich herum ganz leise geworden ist."

## Durchführung 1

„Begib dich in eine bequeme Position ...... Lasse die Augenlider nach unten fallen ...... Sprich deine Entspannungsformel leise vor ...... **‚Ich lass die Muskeln einfach locker ...... Sitz entspannt auf meinem Hocker ...... Atme ein und atme aus ...... Fühl mich wie im Schneckenhaus ...... Schließ die Augen, wenn ich will ...... Plötzlich ist es mäuschenstill!‘** ...... Höre nun in die Stille hinein ...... Kannst du trotz Ruhe noch Geräusche hören?“

## Durchführung 2

„Begib dich in eine bequeme Position ...... Lasse die Augenlider nach unten fallen ...... Sprich deine Entspannungsformel leise vor ...... **‚In meinem Kopf, da geht es zu ...... Es plappern 1 000 Affen ...... Ich bring die Tierchen kurz zur Ruh ...... Ich weiß, ich kann das schaffen ...... Ich atme still, das Denken ruht ...... Ich spüre schon, wie gut das tut.‘** Höre nun in die Stille hinein ...... Kannst du trotz Ruhe noch Geräusche hören?“

## Rücknahme

„Beende die kurze Entspannung mit einem Gähnen, Strecken und Dehnen. Schüttel deinen Körper wohltuend aus.“

## Abschluss

„Welche Geräusche konntest du trotz der Stille noch hören?“

# Die Jahreszeiten

## Ziel der Übung

- Die Übung wirkt gedankenberuhigend und ruhefördernd.

## Das sollten Sie wissen

- Thematisieren Sie die Jahreszeiten. Die Übung eignet sich u. a. für den Musikunterricht.
- Sie können auch Vivaldis „Die vier Jahreszeiten" als Hintergrundmusik abspielen.

## Mentale Einstimmung

„Der Komponist Antonio Vivaldi hat sich vor vielen Jahren überlegt, wie man die Jahreszeiten mit Melodien beschreiben kann. Wie erkennst du die Jahreszeiten? Jede Jahreszeit hat bestimmte Feste, welche kennst du? Probiere es wie Vivaldi und versetze dich in die momentane Jahreszeit. Welche Bilder und Gedanken hast du?"

## Durchführung

„Mache es dir so bequem wie möglich ...... Schließe deine Augen oder senke deine Augenlider ...... Atme tief ein ...... Und wieder aus ...... Kennst du unsere momentane Jahreszeit ...... Ist es gerade Frühling ...... Sommer ...... Herbst ...... oder Winter? ...... Woran erkennst du die Jahreszeit? ...... An der Temperatur? ...... An der Kleidung? ...... An der Natur? ...... Am Stand der Sonne? ...... Stelle dir vor, es wäre deine Lieblings-Jahreszeit ...... Was würdest du dann nach der Schule machen? ......"

## Rücknahme

„Öffne deine Augen ...... Dehne und strecke dich ...... Schüttel dich kräftig aus ......"

## Abschluss

„Welche Jahreszeit ist gerade? Wie fühlt sie sich an? Hast du eine Lieblings-Jahreszeit?"

# Der Atemfluss

## Ziel der Übung

- Diese Übung eignet sich als **Stresslöser** vor und nach langer geistiger Anstrengung und beruhigt gezielt Körper und Geist durch eine bewusste Atemlenkung.

## Das sollten Sie wissen

- **Frische Luft** oder ein angenehmer Zitrusduft unterstützen das Atmen und die Entspannungsmotivation.
- Mit einer **aufrechten** Steh- oder Sitzposition oder einer flachen Liegeposition fällt das Atemnachspüren leichter.

## Mentale Einstimmung

„Unser Körper benötigt ständig Sauerstoff, z. B. zum Denken, zum Verdauen unserer Mahlzeiten oder einfach nur, damit er gesund bleibt. Hast du dich auch schon mal gefragt, wie der Sauerstoff in unsere Fingerspitzen oder in unsere Füße gelangt? Ganz einfach! Durch unsere Blutbahnen fließt der Sauerstoff im Blut überall hin, wo er gebraucht wird. Er fließt bis zur kleinsten Zehe nach unten und auch bis zum höchsten Punkt unseres Kopfes. Versuche doch mal, dein Einatmen gedanklich zu beobachten. Schaffst du es, den Atemfluss zu verfolgen?“

## Durchführung

„Bringe deinen Körper in eine aufrechte Sitzposition ...... Schließe deine Augen ...... Spüre den Einatem ...... und den Ausatem ...... Atme wieder ein ...... Dein Atem versorgt deinen Lieblingsarm ...... deinen Oberarm ...... den Ellenbogen ...... den Unterarm ...... deine Lieblingshand. Beobachte den nächsten Einatem ...... Er fließt über den anderen Oberarm ...... in den Ellenbogen ...... in den Unterarm ...... in die Hand ....... Nimm den nächsten Atemzug ...... Er fließt durch deinen Bauch ...... in dein Lieblingsbein ...... in den Oberschenkel ...... durch dein Knie ...... durch den Unterschenkel ...... in den Fuß ....... Dein nächster Einatem fließt durch den Bauch

..... in den anderen Oberschenkel ...... durch das Knie ...... in den Unterschenkel ..... in deinen anderen Fuß ....... Beobachte deinen nächsten Einatem ...... Er versorgt auch deinen Kopf ...... dein Gesicht ...... die Augen ...... die Nase ...... den Mund ...... den Hals .......“

## Rücknahme

„Öffne die Augen und schüttel deinen Körper wieder munter ...... Atme in deinem Tempo weiter .......“

## Abschluss

„Hast du den Atem achtsam verfolgen können?“

# Die stille Minute

## Ziel der Übung

- Bauen Sie die stille Minute häufig in Ihren Unterricht ein, damit sich die Schüler **an das achtsame Stille-Ritual**, z. B. bei großer Unruhe, **gewöhnen**.

## Das sollten Sie wissen

- Um die **Entspannungsbereitschaft und die Fähigkeit, die Zeit achtsam wahrzunehmen,** zu **steigern**, wiederholen Sie die Übung beliebig oft oder verlängern Sie sie.
- Lassen Sie die Kinder auch vom Sitzen ins Stehen kommen.
- Geben Sie eine **positive Rückmeldung.** Wer hat die Minute exakt erraten?

## Mentale Einstimmung

„Wie lange dauert eine Minute? Ob für uns eine Minute lang oder kurz dauert, hängt davon ab, ob wir sie z. B. interessant oder langweilig empfinden. Beim Spielen vergeht die Zeit schneller als beim Hausaufgabenmachen, stimmt's? Aber wie lange dauert eine Minute, wenn du dich auf sie konzentrierst? Zähle mit deiner Klasse von 5 bis 0 und schätze ab, wie lange deine Minute dauert. Ist sie vorbei, setze dich ganz leise auf deinen Stuhl, damit deine Mitschüler nicht hören, wann deine Minute vorbei ist."

## Durchführung

„Stelle dich entspannt auf deinen Platz ...... Lege die Augenlider sanft nach unten ...... Atme tief ein und wieder aus ...... 5– 4 – 3 – 2 – 1 – 0 ...... Die Minute läuft."

## Rücknahme

„Öffne deine Augen ...... Schüttel dich aus und strecke deine Arme in alle Richtungen."

## Abschluss

„War es leicht oder schwer, eine Minute zu bestimmen? Hast du mitgezählt?

# Das innere Lächeln

## Ziel der Übung

- Diese Übung **lockert die Muskulatur, löst Anspannungen,** z. B. vor oder nach Klassenarbeiten, und hebt die Stimmung.

## Das sollten Sie wissen

- Leise Musikuntermalung **unterstützt** die **Entspannungsbereitschaft** der Kinder.
- Mit auf dem Tisch abgelegtem Oberkörper vermeiden Sie evtl. Störungen.
- Seien Sie so oft es geht als **lächelnder Lehrer ein Vorbild** für Ihre Schüler.

## Mentale Einstimmung

„Wann hast du das letzte Mal so richtig gelacht? Wusstest du, dass Kinder viel häufiger lachen als Erwachsene? Warum ist das so? Es gibt Lachforscher, die herausgefunden haben, dass das Lachen, egal ob herzhaft oder grundlos, Glückshormone in unseren Körper zaubert und wir uns glücklicher und entspannter fühlen. Haben die Lachforscher Recht?"

## Durchführung

„Nimm eine entspannte Sitzhaltung ein ...... Schließe oder fokussiere deine Augen ...... Zaubere ein Lächeln auf deine Lippen ...... Atme dein Lächeln tief ein ...... Lenke es vom Kopf über die Arme in die Fingerspitzen ...... Atme noch ein Lächeln ein ...... Lenke es zum Herzen ...... Atme dein Lächeln ein ...... Lenke es in den Bauch ...... die Beine ...... die Zehenspitzen ...... Spüre das Lächeln im ganzen Körper ......"

## Rücknahme

„Öffne deine Augen ...... Strecke dich nach allen Seiten ...... Lächle leise in dich hinein."

## Abschluss

„Konntest du das Lächeln spüren? Wo hast du es am stärksten gespürt?"

# Der Greif-Profi

## Ziel der Übung

- Diese Übung eignet sich als Ruheübung vor Klassenarbeiten und zur Wahrnehmungsschulung.

## Das sollten Sie wissen

- Geben Sie den Kindern Zeit, ihre Empfindungen wahrzunehmen und den Gegenstand zu erspüren.
- Die achtsame Wahrnehmung lässt sich nach kurzer Übungszeit auch auf andere Bereiche übertragen, z. B. das achtsame Wahrnehmen von Gefühlen und Stimmungen in der Klasse, auf dem Pausenhof, vor und nach der Klassenarbeit.

## Mentale Einstimmung

„Wie viele Sinne hat der Mensch? Kannst du sie alle benennen? Richtig! Ein gesunder Mensch hat fünf Sinne. Er kann sehen, hören, riechen, tasten und schmecken. Leider gibt es auch Menschen, denen ein Sinn fehlt, z. B. bei einem blinden Menschen. Damit sich Blinde im Alltag besser zurechtfinden können, haben sich die anderen Sinne stärker entwickelt. Blinde lernen, mit den Ohren und den Fingern zu ‚sehen'. Der Tastsinn, der Hörsinn, der Geschmackssinn und der Geruchssinn helfen dem Blinden, seine Umwelt, z. B. Gegenstände und Lebensmittel, besser zu erkennen. Willst du mal testen, wie gut du ohne deinen Sehsinn zurechtkommst?"

## Durchführung

„Setze dich entspannt auf deinen Stuhl ...... Schließe deine Augen ...... Hole einen Stift aus deinem Mäppchen ...... Erkunde deinen Stift mit deinen Fingern ...... Wie fühlt sich die Oberfläche an? Glatt oder rau? ...... Wie ist die Form? Eckig oder rund? ...... Wie lang ist der Stift? Länger oder kürzer als deine Zeigefinger? ...... Wie ist die Spitze? Angenehm weich oder unangenehm spitz? ...... Hat dein Stift einen Radiergummi? Wonach riecht dein Stift? ...... Aus welchem Material ist dein Stift?"

## Rücknahme

„Öffne nun deine Augen ...... Schaue dir deinen Stift noch mal genau an ...... Jetzt hast du deinen Stift mit deinen Sinnen erkundet."

## Abschluss

„Würdest du diesen Stift nun auch mit geschlossenen Augen in deinem Federmäppchen wiedererkennen? Probiere es doch mal aus."

# Das Mühlrad

## Ziel der Übung

- Achtsamkeit auf die **Lockerung der Schulter- und Nackenmuskulatur**

## Das sollten Sie wissen

- Singen Sie mit den Schülern den Anfang des Liedes „Es klappert die Mühle am rauschenden Bach“. Das unterstützt einen **entspannten, langsamen Atemrhythmus.**
- Zur **Intensivierung der Muskelübung** können die „Mühlenarme“ einzeln „mahlen“, die Augen heften sich dabei an die Fingerspitzen.

## Mentale Einstimmung

„Hast du schon mal einem alten Mühlrad beim Drehen oder Mahlen des Getreides zugeschaut? Es gibt sie noch selten und meist als Attraktion für Touristen.“

## Durchführung

„Stelle dich an deinen Platz ...... Deine Arme hängen nach unten ...... Stelle dir ein Mühlrad vor ...... Die Kraft kommt aus der Achse des Mühlrads ...... deiner Körpermitte ...... Bewege es mit deinem Atem und deinen Armen ...... Strecke Arme und Finger weit nach außen ...... Beide Arme kreisen gleichmäßig ...... Dein Mühlrad bewegt sich mit deinem Atem ...... Atme langsam ein und aus ...... Bringe das Rad zum Stehen ...... Beide Arme hängen nach unten ...... Bewege das Rad in die andere Richtung.“

## Rücknahme

„Lasse nun das Mühlrad langsam zur Ruhe kommen ...... Deine Armmuskeln sind ganz schwer und entspannt ...... Schüttel deine Muskeln wieder locker.“

## Abschluss

„Kannst du jetzt Muskeln deutlicher spüren als vorher? Welche?“

# Das Mühlrad

# Der Morgenstarter

## Ziel der Übung

- Regelmäßig durchgeführt, kann der Morgenstarter die **Wahrnehmung der eigenen Befindlichkeit** deutlich verbessern und **positiv steuern**. Dies kann wiederum die Gesamtstimmung einer Klasse positiv beeinflussen und bei der Aufnahme und Verarbeitung des Lernstoffes unterstützen.

## Das sollten Sie wissen

- Kinder starten je nach familiären Gegebenheiten und Vorkommnissen mit ganz unterschiedlichen Gefühlen und Gedanken in den Schultag. Ein **morgendliches Ankommritual** kann den Schülern die Einstimmung auf das Lernen erleichtern. Beginnen Sie den Schultag deshalb so oft Sie können mit einem **positiven Morgenstarter.**

## Mentale Einstimmung

„Kaum bist du morgens aufgewacht, schon bestimmen deine Gedanken, wie du dich fühlst. Mit welchen Gedanken und Gefühlen du morgens aufwachst, hängt meist damit zusammen, wie dein Körper sich im Schlaf erholen konnte. Böse Träume oder ein zu kurzer Schlaf können darüber entscheiden, ob du gut gelaunt oder schlecht gelaunt aufwachst. Ob du dich den ganzen Tag gut oder schlecht gelaunt fühlen möchtest, bestimmst heute mal ganz alleine du, indem du deine Gedanken steuerst. Heute bist du ein Gefühle-Detektiv auf der Suche nach angenehmen oder weniger angenehmen Gefühlen in deinem Körper. Beobachte bei der folgenden Übung, welche Gefühle sich momentan in deinem Körper befinden, und bestimme anschließend, dich heute wohlzufühlen."

## Durchführung

„Lass deinen Körper ganz entspannt und locker in deinen Stuhl sinken ...... Deine Augenlider fallen entspannt nach unten ...... Beobachte deine Atemzüge ...... Beobachte deine Gedanken ...... Woran denkst du? ...... Beobachte, wie sich dein Körper anfühlt ...... Gibt es eine Stelle, die du ganz deutlich spüren kannst? ...... Vielleicht im Bauch ...... oder im Kopf ...... oder aber im ganzen Körper? ...... Heute bestimmst du, wie sich dein Körper anfühlen soll ...... Lenke mit der folgenden Zauberformel ein gutes Gefühl in deinen Körper ...... Atme ein ...... **‚Mir geht es gut'** ...... Atme aus ...... **‚Ich fühle mich wohl'** ...... Atme ein ...... **‚Mir geht es gut'** ...... Atme aus ...... **‚Ich fühle mich wohl'** ...... Atme ein ...... **‚Mir geht es gut'** ...... Atme aus ...... **‚Ich fühle mich wohl'** ...... Genieße das Wohlfühlgefühl noch eine Weile ......"

## Rücknahme

„Wecke nun deine Muskeln wieder auf ...... Strecke und dehne deine Arme und Beine ...... Schüttel sie aus ...... Öffne deine Augen ...... Spüre noch mal kurz dein gutes Gefühl."

## Abschluss

„Wo spürst du in deinem Körper am deutlichsten, wie es dir im Augenblick geht?"

# Das Überkreuzatmen

## Ziel der Übung

- Diese kinesiologische Übung **löst Blockaden und aktiviert den Energiefluss.**

## Das sollten Sie wissen

- Das Überkreuzatmen kann auch **im Liegen und Sitzen** gut durchgeführt werden.
- Führen Sie die Übung täglich durch, um bei den Kindern schon nach wenigen Tagen eine **spürbar bessere Entspannungsbereitschaft** festzustellen.

## Mentale Einstimmung

„Wusstest du, dass du deinen Atem durch den Körper lenken kannst? Gedanklich zauberst du dem Atem den Weg frei, damit er durch deine Blutbahnen fließt. Probiere es aus!"

## Durchführung

„Stelle dich gut geerdet an deinen Platz ...... Strecke beide Arme nach oben ...... Atme 3-mal tief ein ...... Dein Bauch wächst nach vorn ...... Atme aus ...... Dein Bauch wird kleiner ...... Atme über die Fingerspitzen deiner rechten Hand tief ein ...... Dein rechter Arm wird länger ...... Atme über deinen linken Fuß tief aus ...... Dein linker Fuß wird ganz schwer ...... Atme entspannt ein ...... Dein Bauch wird groß ...... Atme aus ...... Dein Bauch wird kleiner......Atme über die Fingerspitzen deiner linken Hand tief ein ...... Dein linker Arm wird länger...... Atme durch deinen rechten Fuß tief aus ...... Dein rechter Fuß wird ganz schwer ...... Atme ruhig und wiederhole die Übung ......"

## Rücknahme

„Lass nun beide Arme fallen und schüttel deinen Körper wieder locker."

## Abschluss

„Konntest du deinen Atem über die Blutbahnen durch deinen Körper lenken?

# Das Überkreuzatmen

# Die Augenreise

## Ziel der Übung

- Mit dieser achtsamen Wahrnehmungsübung lenken Sie die Aufmerksamkeit gezielt und Schritt für Schritt in die **entspannte und reinigende Innenwelt** der Kinder.

## Das sollten Sie wissen

- **Leise Hintergrundmusik** unterstützt angenehm die Ruhebereitschaft der Schüler.
- Um die Aufmerksamkeit leichter nach innen lenken zu können, empfehle ich, die Augen mit den Händen zu verdecken oder den **Kopf auf dem Tisch abzulegen**. Sie können die Übung auch **liegend im Freien** durchführen.

## Mentale Einstimmung

„Unsere Augen sind täglich viele Stunden aktiv und sehen dabei eine Menge Bilder, die unser Gehirn erkennen und verarbeiten muss. Unser Gehirn entscheidet dabei blitzschnell, welche Bilder wichtig sind und im Unterbewusstsein gespeichert werden und welche gleich wieder gelöscht werden. Manchmal ist der Speicher auch sehr voll. Dann freut er sich über viel Ruhe und Entspannung. In dieser Erholungszeit werden alle unwichtigen Informationen gelöscht. Versuche doch mal, herauszufinden, welche Bilder dein Gehirn speichert und wie viel Platz noch in deinem Speicher vorhanden ist."

## Durchführung

„Richte deinen Blick aus dem Fenster ...... Bringe deinen Körper in eine entspannte Lage ...... Atme ein paar Atemzüge bewusst ein und aus ...... Schicke deine Augen auf eine Reise durch das Fenster ...... Was kannst du alles entdecken ...... Beobachte alles, was für dich wichtig ist ...... Lege nun deinen Kopf auf dem Tisch ab ...... Senke deine Augenlider ...... Lenke deinen Blick nach innen ...... Welche Bilder tauchen in dir auf? ......"

## Rücknahme

„Öffne nun wieder deine Augen ...... Richte deinen Körper wieder auf ...... Dehne, strecke und schüttel dich aus ......"

## Abschluss

„Welche Bilder hat dein Gehirn gespeichert? Gibt es Zeiten, an denen dein Speicher übervoll ist?"

# Der Zauberatem

## Ziel der Übung

- Die Übung eignet sich zum Durchatmen **nach anstrengenden Tätigkeiten.**

## Das sollten Sie wissen

- Diese Übung lenkt die Achtsamkeit zur Atmung und verdeutlicht die Macht unserer **Vorstellungskraft**, unserer Gedanken.

## Mentale Einstimmung

„Wusstest du, dass du mit deinem Atem Gegenstände größer und kleiner zaubern kannst? Dazu brauchst du nur Nase oder Mund und eine gute Fantasie."

## Durchführung

„Stelle dich an deinen Platz …… Deine Arme hängen herab …… Schließe die Augen …… Stelle dir einen Stecknadelkopf vor …… Atme deinen Stecknadelkopf so groß wie einen Apfel …… Stelle dir nun eine Melone vor …… Atme deinen Apfel so groß wie eine Melone …… Stelle dir nun einen Sitzball vor …… Atme deine Melone so groß wie einen Sitzball …… Stelle dir nun einen Heißluftballon vor …… Atme deinen Sitzball so groß wie einen Heißluftballon …… Stelle dir nun einen Fußball vor und lasse mit deinem Ausatem den Heißluftballon klein werden wie einen Fußball …… Stelle dir nun eine Mandarine vor …… Atme aus und lasse den Fußball klein werden wie eine Mandarine …… Stelle dir einen Punkt vor …… Atme aus und lasse die Mandarine klein werden wie einen Punkt."

## Rücknahme

„Öffne deine Augen und schüttel dich im Stehen gut aus."

## Abschluss

„Konntest du die Gegenstände wachsen und schrumpfen lassen?"

# Der Gedankenfischer

## Ziel der Übung

Wohltuende und effektive **Gedankenberuhigung** vor schulischen Tests

## Das sollten Sie wissen

**Vermeiden Sie** Entspannungsmusik mit **Wasser- oder Meeresrauschen.** Diese aktiviert häufig den Harndrang der Schüler und bringt oft Unruhe ins Klassenzimmer.

## Mentale Einstimmung

„Täglich denken wir viele Tausend Gedanken. Mal gute, mal weniger gute Gedanken. Hast du schon mal versucht, deine schlechten Gedanken zu fangen? Heute darfst du ein Fischer sein, der sein Netz ausbreitet, um alle schlechten Gedanken zu fangen und sie weit weg ins Meer zu spülen. Probiere es doch mal aus!"

## Durchführung

„Setze dich bequem auf deinen Stuhl ...... Senke deine Augenlider ...... Beobachte deine Atemzüge ...... Stelle dir nun ein Fischernetz vor ...... Es ist sehr groß und hat enge Maschen ...... Auch ganz kleine Gedanken, die du loswerden willst, bleiben darin hängen ...... Breite das Netz aus ...... Fange alle Gedanken, die du loswerden willst ...... (lange Pause) ...... Nun ist dein Netz sehr voll ...... Lasse es los ...... Beobachte, wie alle Gedanken von den Wellen weit weggespült werden ......"

## Rücknahme

„Komme wieder aus deiner Meereswelt in unseren Klassenraum zurück ...... Schüttel dich aus ...... Dehne deine Arm- und Beinmuskeln ...... Öffne langsam deine Augen."

## Abschluss

„Hattest du viele Gedanken im Netz?"